Phenomenalism

Phenomenalism

A Metaphysics of Chance and Experience

MICHAEL PELCZAR

OXFORD
UNIVERSITY PRESS

OXFORD
UNIVERSITY PRESS

Great Clarendon Street, Oxford, OX2 6DP,
United Kingdom

Oxford University Press is a department of the University of Oxford.
It furthers the University's objective of excellence in research, scholarship,
and education by publishing worldwide. Oxford is a registered trade mark of
Oxford University Press in the UK and in certain other countries

First Edition published in 2023

Impression: 1

Published in the United States of America by Oxford University Press
198 Madison Avenue, New York, NY 10016, United States of America

British Library Cataloguing in Publication Data
Data available

Library of Congress Control Number: 2022946074

ISBN 978–0–19–286873–2

DOI: 10.1093/oso/9780192868732.001.0001

Printed and bound in the UK by
Clays Ltd, Elcograf S.p.A.

If it looks like a duck, swims like a duck, and quacks like a duck, then it is probably a duck.

—unknown

Contents

Acknowledgements

I began work on this book in 2016, during a sabbatical leave at Princeton, whose hospitality I gratefully acknowledge. Since then, the book has gone through many changes, including several complete tear-downs and rebuilds. Throughout the process, I've had the benefit of excellent advice and sharp but always good-natured criticism from a variety of sources. Geoffrey Lee, Eugene Mills, and Kevin Morris contributed to a symposium on my first book about phenomenalism, organized by Neil Manson; thinking through their comments was the first step towards the present book. My colleagues in Singapore were supportive and insightful as always; I should particularly mention my current colleagues Zach Barnett, Bob Beddor, Ben Blumson, Ethan Jerzak, Lavinia Picollo, Abelard Podgorsky, Qu Hsueh Ming, Neil Sinhababu, Tang Weng Hong, and Dan Waxman, and past colleagues Fatema Amijee and Frank Jackson. Howard Robinson provided detailed and incisive comments on a penultimate draft of the manuscript, which resulted in many improvements (though none, I expect, that will bring him over to the Millian side of mind-first metaphysics). In 2017, David Chalmers organized an idealism workshop in Shanghai, where I presented some ideas related to the book; I received excellent feedback from the workshop participants on that occasion. Also in 2017, I presented a paper on phenomenalism at Chapel Hill North Carolina, where I again received excellent feedback, especially from Rob Smithson and Thomas Hofweber. In 2018 I presented a different phenomenalism-related paper as part of an idealism symposium organized by Craig Warmke for the Central APA meeting in Chicago that year; I got great input from the attendees of that symposium, and especially from Brian Cutter, who commented on my paper. Helen Yetter-Chappell and her graduate students at the University of Miami read a complete draft of the book in 2019, and their comments prompted many changes. I should also thank Andrew Bailey, John Burgess, Stuart Derbyshire, Alan Hájek, Mark Johnston, Robert Kirk, Uriah Kriegel, Harold Langsam, Neil Mehta, David Papineau, and Galen Strawson for their thoughts on themes directly related to the book, and last but not least the anonymous readers for Oxford University Press for their valuable comments and suggestions.

M.W.P.

Preface

Ask a metaphysician today about the relationship between mind and matter, and you'll probably hear a lot about whether consciousness reduces to something physical, like a brain state. The unspoken assumption is that the physical world is familiar, well-understood, and ultimately unmysterious. The question is what to make of conscious experience, including the experience that reveals the world to us.

Travel back in time a hundred years or so to ask the same question, and you'll hear a lot about whether physical things reduce to something mental. The unspoken assumption was that conscious experience was something familiar, well-understood, and ultimately unmysterious. The question was what to make of the world that experience reveals.

Return to the present, and you might notice signs of renewed interest in the older question. With the prospects for a reductive theory of consciousness looking increasingly dim, and physics looking increasingly incapable of providing its own interpretation, some people are giving reductive theories of the physical a second chance, including theories that attempt to define the physical in mental terms rather than vice versa.[1]

This book is my contribution to the revival of mind-first metaphysics. Its aim is to convince you that physical things are nothing more than tendencies for experiences to occur in certain ways—"permanent possibilities of sensation," as J.S. Mill calls them. Many have found Mill's proposal intriguing, but it's not easy to turn it into a proper theory. What does "possibility" mean in this context, and what sort of experiences are physical things supposed to be possibilities for? Historic treatments of phenomenalism leave these questions largely unanswered. In the pages that follow I try to do better.

[1] Prominent examples include David Chalmers, Rae Langton, and Galen Strawson: see Chalmers (1996: 297–308), Langton (1998), and Strawson (2006). Two philosophers never stopped asking the older question: see Foster, (1982; 1991; 2008) and Robinson (1982, 1994, forthcoming).

At the end of the day, phenomenalism is a simple idea: a rock is a tendency for experiences to occur as they do when people perceive a rock, and likewise for all other physical things. This book is an elaboration of that idea.

M.W.P.

Singapore
February 2021

List of Figures

1
The World as Hypertext

There's a cantaloupe on the table. This has various implications. For one, if you were looking at the table, you'd have a visual impression of a roundish object with a variegated brownish green surface. For another, if you were to touch the cantaloupe, you'd have tactile experiences of something cool and hard. More generally, there's a complex package of experiences that a person appropriately situated with respect to the cantaloupe is apt to have. This package of possible experiences is arguably unique. An object ever-so-slightly different from this particular melon would reveal its presence through slightly different experiences, at least in appropriately situated observers. It would have a slightly different phenomenal fingerprint, so to speak. A completely dissimilar object would come with a very different package of possible experiences.

This is all fairly obvious, but it raises an important philosophical question about the relationship between the melon and the associated package of possible experiences.

According to a very natural commonsense picture, the melon is in some sense prior to the possible experiences. The experiences we're apt to have when perceiving the melon depend on the melon, which somehow underlies the experiences or their possibility.

In this book, we're going to explore a different picture of the relationship. Could it be that the possible experiences are in fact fundamental, and that the melon just *is* the complex package of possible experiences? This is a radical departure from the common-sense view: a surprising inversion of our ordinary way of thinking. Still, there might be something to it. After all, we only know of the melon by actually having some of the possible experiences associated with it. We never get any direct evidence of something underlying our experiences of physical things, existing independently of those experiences or the possibilities for them. If we want to keep our ontology as simple as our evidence allows, there's at least something to be said for the inverted way of looking at things.

The standpoint I've described is called phenomenalism. It invites various objections. You might wonder how phenomenalists can make sense

Phenomenalism: A Metaphysics of Chance and Experience. Michael Pelczar, Oxford University Press.
 DOI: 10.1093/oso/9780192868732.003.0001

of possibilities for experience without positing an underlying reality that grounds or sustains the possibilities; you might wonder how phenomenalists can describe the relevant possibilities without calling on observers with appropriately constituted brains and sensory organs; you might wonder how to reconcile phenomenalism with the intersubjectivity of physical things. We'll address these issues and others in due course. As we do so, and the more we reflect on the somewhat revisionary picture that phenomenalism presents, the more natural the view will come to seem—at least, that's my hope. I also hope to show that phenomenalism sheds much-needed light on the connection between everyday experience and modern physics, the relationship between mind and body, and the nature of perceptual experience.

In the phenomenalist worldview, there are experiences, and there are possibilities for experience—and that's all. Later, I'll argue that the possibilities are best understood as conditional probabilities for various experiences to occur given the occurrence of other experiences. So, in my view, reality consists of experiences, and probabilities related to experience. The probabilities are the physical part of the world.

Let's work our way a little deeper inside the phenomenalist worldview. Imagine that, unbeknownst to us, there are conscious beings who perceive without being perceived, and have no effect on the physical world. Imagine that our universe is thickly populated with such beings. Suppose that these beings—call them *ideal observers*—are distributed in time and space in such a way that no physical feature of our world escapes their notice: collectively, they perceive every physical object, event, process, and state of affairs that exists in our world at any time or place.

An earthquake occurs off the coast of Sumatra, causing a tidal wave in Somalia. The occurrence of the wave event depends on the occurrence of the tectonic event. We might put this by saying that the tectonic event gives a certain probability to the occurrence of the wave event, or that the occurrence of the tidal wave counterfactually depends on the tectonic shift. Given the ubiquity of ideal observers, there are experiences of the earthquake, and experiences of the tidal wave, and the latter depend on the former to the same degree that the occurrence of the tidal wave depends on the occurrence of the earthquake. Likewise for all the events that occur between the earthquake and the tidal wave: as a water-wave propagates through the Indian Ocean, an experience-wave propagates through the ideal observers.

What if the only thing we knew about were the ideal observers' experiences? We know about some observers' earthquake experiences, we know about other observers' tidal wave experiences, and we know these experiences stand in various counterfactual or probabilistic relations. We have, let's suppose, complete information about the phenomenal qualities of all the observers' mental lives, and complete information about the dependencies among the observers' experiences. What should we infer from this information?

You might think it would be appropriate to infer the existence of an independently existing physical world with features that give the ideal observers their experiences and explain the dependencies among them.

But there's another, more conservative, answer. What should we infer from the given information? Nothing! What calls for explanation about the ideal observers' experience is its apparent non-randomness: the observers' experiences occur in regular and seemingly coordinated ways. It's reasonable to think that something explains this fact. The minimum required to explain it are objective tendencies or propensities for the observers to have certain experiences, conditional on their having certain other experiences. But we've already inferred these tendencies, by positing relations of probabilistic or counterfactual dependence among the ideal observers' experiences. To explain the propagation of the experience-wave, it's enough to suppose that certain observers have experiences as of rising water, which leads to certain other observers having experiences as of falling water, which leads to yet other observers having experiences as of rising water, and so forth. There's no obvious need to posit anything further to explain the patterns in the observers' experiences.

If this sounds like an invitation to external world skepticism, that's not how it's intended. As I see it, once we've concluded that certain experiences are apt to occur in certain ways conditional on the occurrence of certain other experiences, we've already arrived at an external world. Physical things just *are* propensities for experiences to occur in certain ways. They are, as Mill puts it, "permanent possibilities of sensation."

The ideal observers are a fiction, but it's at least possible for there to be experiences just like theirs, related to one another in just the same ways. For any possible physical world, we can imagine a network of experiences that would exist if that world were thickly populated with ideal observers. Call any such network of experiences an *ideal world.*

Given that our world doesn't contain anything like the number and variety of experiences that exist in a scenario with ideal observers, the physical world we inhabit is not an ideal world. Still, we might identify physical reality with the circumstance that a certain ideal world is the one that *would* exist, if there were an ideal world; or, the circumstance that a certain ideal world is the one that would exist, if our own actual experiences were parts of an ideal world; or, the circumstance that a certain ideal world is the one with the highest probability of existing, conditional on our own experiences belonging to an ideal world. Phenomenalism is the view that physical reality is some such circumstance.

According to phenomenalists, physical facts are conditional facts about experience. Exactly what kind of conditional facts is negotiable: traditional phenomenalists think of them as counterfacts of the form, "If experiential state of affairs ϕ existed, experiential state of affairs ψ would exist"; I prefer to think of them as probabilistic facts of the form, "The likelihood of experiential state of affairs ϕ given experiential state of affairs ψ = (or >, or <) x." But all phenomenalists agree that for there to be some physical entity is just for there to exist some conditional state of affairs related to experience.

Loosely following Mill, let's use "possibilities of sensation," "possibilities for experience," and "phenomenal possibilities" synonymously, as blanket terms for conditional facts about experience.[1]

Phenomenalists think that possibilities of sensation are enough to account for the prima facie orderliness and intersubjective coordination of actual experiences. There's no need to posit a further level of reality that grounds or underlies the possibilities. Phenomenalists do however acknowledge that there's a reason why you have the experiences you do. You have them because various physical events take place in your brain, which are themselves effects of further events in your physical environment. But your brain and environment are just possibilities of sensation: the physical facts about them are conditional facts about experience. Some phenomenal possibilities explain others, and some explain actual experiences, but there's no need to posit something over and above all phenomenal possibilities to explain our experiences or anything else.

It's reasonable to think that something explains why experiences occur in the seemingly coordinated and non-random ways they do, and phenomenalists say that what explains it are various conditional facts about experience. Opponents of phenomenalism think that the conditional facts

[1] I hope it's clear that these are terms of art. The phrases "possibility of sensation," "possibility for experience," and "phenomenal possibility" are just convenient ways of referring to conditional facts about experience, without committing ourselves to any particular view about the exact nature of the relevant conditionals (e.g., counterfactual vs. probabilistic). Needless to say, the "possibility" in "possibility of sensation" doesn't mean mere logical possibility.

about experience are grounded in an underlying reality, and that it's this underlying reality that is the physical world.

Well, suppose we do posit an underlying reality: something that doesn't reduce to conditional facts about experience, and that grounds or underlies the conditional facts about experience. What does this achieve?

Nothing, without more details about the nature of the posited reality and the mechanism by which it ostensibly grounds or underlies the phenomenal possibilities that characterize our world. However, if something grounds or underlies the phenomenal possibilities, we can't learn anything about it except through experience. If you try to look behind experience to see where it's coming from, you're just going to have more experience.

So, if there's something distinct from all experiences and possibilities for experience that explains why they exist, we can never know anything about it, except possibly (but how?) that it explains why the experiences and possibilities for experience exist. Positing such an entity is like saying that something causes the tide to rise and fall, adding that the only thing we can ever know about this "something" is that it causes the tide's rise and fall.

Maybe there is something that grounds all the conditional facts about experience that hold in our world; at least, this isn't something phenomenalists have to deny. But there's no compelling reason to think there is any such grounding entity, and even if there is, the physical world is not it, but the conditional facts about experience that it grounds. The way phenomenalists see it, if something grounds the conditional facts about experience, it explains why there is a physical world, but it is not itself the physical world: its relationship to the physical world is analogous to God's in traditional theistic cosmology.

Phenomenalists see the physical world as analogous to a hypertext. The melon on the table is a collection of hyperlinks: click for visual sensations of something round and beige; click again for tactile sensations of something cool and rough; click once more for visual sensations of something fleshy and salmon-colored. With luck, the next click will take you to some sweet and juicy flavor-sensations. (Hopefully there won't be any annoying pop-ups, like a fly landing on your plate.)

The hypertext analogy only goes so far. In an actual hypertext, like the World Wide Web, links exist only as suitably configured computer hardware: take away the underlying hardware, and you take away the hypertext. In the phenomenalist view, nothing underlies a hyperlink, except possibly more hyperlinks. The computer hardware itself is just a network of hyperlinks. It's hyperlinks all the way down.

Phenomenalism is such an uncommon view today that I should say something about how it relates to more familiar metaphysical theories, like dualism, materialism, and traditional idealism.

Like idealism and materialism, but unlike dualism, phenomenalism is a kind of mind–body monism. Materialism reduces all mental facts to physical facts; phenomenalism reduces all physical facts to mental facts. It's true that the mental facts to which phenomenalists reduce physical facts aren't facts about actually occurring experiences (as in traditional idealism), but about what experiences occur conditional on the occurrence of other possible experiences. But this doesn't disqualify phenomenalism from counting as a kind of monism. Materialist theories typically include in their ontology various conditional facts about physical phenomena. If conditional facts about experience are non-mental, it's only in the same sense that conditional facts about the physical world are non-physical; so, if typical materialist theories are monistic, so is phenomenalism.[2]

Like dualism and idealism, but unlike materialism, phenomenalism doesn't entail a reductive view of consciousness. Phenomenalists identify all physical things with possibilities for experiences, but they do not, or anyway need not, identify any experience with anything physical. In particular, phenomenalists are under no obligation to identify conscious experiences with brain states. Phenomenalism is compatible with the possibility of zombies.[3]

Like dualism and materialism, but unlike traditional idealism, phenomenalism doesn't entail that physical things are mind-dependent. That's

[2] My policy will be to classify conditional facts about experience as phenomenal facts, and conditional facts about physical phenomena as physical facts. (So, in my taxonomy, the phenomenal facts include both facts about actually occurring experiences and conditional facts about experience, and the physical facts include both facts about actually occurring physical phenomena and conditional facts about physical phenomena.) This is just a terminological choice. If we classify conditional facts as topic-neutral instead of phenomenal or physical, phenomenalism and materialism will be dualistic, positing topic-neutral facts in addition to facts about occurrent mental or physical phenomena; but then what's normally called dualism will posit three kinds of facts: occurrent physical, occurrent mental, and conditional.

[3] Phenomenalism is also compatible with the claim that zombies are impossible. You could combine the view that physical entities are possibilities of sensation with the view that conscious states are brain states; you'd just have to add that the brain states are nothing but possibilities of sensation. (Something like this seems to be Hobbes's view: see Hobbes 1655/1839: 102–4, 387–93.) Such a theory would inherit all the vulnerabilities of standard materialism—the zombie argument, the knowledge argument, etc.—and for that reason doesn't strike me as an attractive position. (Zombie argument: there could be possibilities for experiences of my brain states in a world with no conscious experience; knowledge argument: Mary could know all the possibilities for experiences of my brain, eyes, etc. without knowing what it's like to see red.) I'll say more about materialistic phenomenalism in Ch. 8.

because, unlike traditional idealists, phenomenalists identify physical things with possibilities for experience, rather than actual experiences. Since a possibility can exist without being realized, a conditional fact about experience can hold even if none of the experiences it concerns actually occurs—phenomenalism allows that rocks could exist in a world devoid of experience.[4]

These comparisons cast phenomenalism in a rather favorable light. It's the only monistic theory among those considered that delivers a world of mind-independent physical things, without committing its proponents to a reductive theory of consciousness.

In some ways, the position closest to phenomenalism is another relatively unknown theory: structuralism (also known as ontic structural realism). In one standard version, structuralism identifies the physical facts of our world with facts about causal structure, understood in terms of causal powers. The difference is that where structuralists want to construct the physical world out of powers to cause nothing but further powers, phenomenalists want to construct it out of possibilities for something besides further possibilities, namely conscious experience. In both views, the physical world is a kind of bare input–output architecture. It's just that for phenomenalists, the inputs and outputs are possible experiences, while for structuralists they're further bits of input–output architecture.[5]

Structuralism is a very austere view: too austere for most tastes. I find it rather attractive, and more defensible than most people think; I don't know of any knock-down argument against it. But structuralism tells us nothing about the relationship between mind and matter, nothing about the relationship between the world as described by physics and the world as revealed in everyday experience, and nothing about the relationship between perception and its objects. Phenomenalism has a clear and (to my mind) satisfying story to tell about all these relationships. Structuralism gives us an analysis of matter; phenomenalism gives us that and much more.

[4] Thus David Armstrong is mistaken when he claims that "unobserved physical reality cannot, for the phenomenalist, be what we all think it is in our unphilosophical moments: something ontologically additional to observed physical reality" (Armstrong 2004: 2) A better target for Armstrong's criticism is Berkeley, who holds that "all those bodies which compose the mighty frame of the world, have not any subsistence without a mind; that their *being* is to be perceived or known; that consequently so long as they are not actually perceived by me, or do not exist in my mind, or that of any other created spirit, they must either have no existence at all, or else subsist in the mind of some Eternal Spirit" (Berkeley 1710/1901: §6).

[5] For structuralism, see Russell (1927), Dirac (1938–9), Ladyman and Ross (2007: 130—89), Sider (2011), French (2014), and Tegmark (2014). More on structuralism in Ch. 7.

The plan for the book is as follows. Chapter 2 presents Mill's original phenomenalist theory, to provide some historical context for what follows. Chapter 3 elaborates the argument for phenomenalism that I sketched earlier in this chapter, which parallels an influential argument for materialism due to David Lewis and David Armstrong. At this stage, I'll be working with an intuitive notion of possibilities of sensation as propensities that support relevant experience-related counterfactual conditionals or conditional probabilities. In Chapters 4 and 5, I make this notion precise: Chapter 4 offers a detailed account of the experiential states of affairs that I take physical things to be possibilities for, and Chapter 5 offers a detailed explanation of what I mean by saying that there are possibilities for them. Chapter 6 gives a second argument for phenomenalism, which parallels another influential argument for materialism, this one due to U.T. Place and J.J.C. Smart. Chapter 7 uses phenomenalism to shed light on some central questions in the philosophy of science. Chapter 8 explores phenomenalism's implications for the mind–body problem. Chapter 9 offers a phenomenalist theory of perception. Chapter 10 wraps things up with a review of my overall position, and a comparison of phenomenalism to David Lewis's Humean supervenience.

Following is a more detailed overview.

Chapter 2 introduces Mill's theory and puts it into historical context, in relation both to the 18th- and early 19th-century metaphysics that preceded it and to the late 19th- to early 20th-century metaphysics that followed. The goal of this chapter is to develop Mill's position in enough detail to make its pros and cons evident. This will make it easier to see what it will take to put phenomenalism back on the map, which is the goal of this book.

Roughly, phenomenalism is the view that the physical world is a potential for things to be as traditional idealists, like Berkeley and Leibniz, think things actually are: hence the popular caricature of phenomenalism as "Berkeley without God." Chapter 2 traces the roots of Mill's phenomenalism back to traditional idealism, explaining how Mill's position grew out of earlier idealist thinking via the Kantian metaphysics that predominated in Mill's day.[6]

Millian phenomenalism played an important role in late 19th- to early 20th-century metaphysics, and philosophers took it seriously into the 1960s. Since then, phenomenalism hasn't received much attention, and the attention it has received has often done more to obscure than illuminate Mill's position. After giving my own interpretation of Mill's theory, I criticize

[6] The "Berkeley without God" caricature is due to G.J. Warnock (1969: 225). As we'll see in Ch. 2, a better caricature would be "Kant without noumena."

several interpretations that cast the theory in an unflattering light. This will give me a chance to dispel some common misconceptions about phenomenalism.[7]

In **Chapter 3**, I develop the argument for phenomenalism that I sketched at the beginning of this chapter: the physical world is whatever explains why our perceptual experiences exhibit the regularities they do; the explanation is that there are objective tendencies for experiences to exhibit those regularities; therefore, the physical things we perceive are such tendencies. I argue that our everyday ways of thinking and acting show that we're already committed to the first premise of this argument. In support of the second premise, I argue that a phenomenalist explanation of the regularity of experience is superior to alternative explanations, such as those we get from Kantians and metaphysical realists.

The purpose of **Chapter 4** is to say exactly what kind of experiential phenomena I take physical things to be possibilities for. Roughly, the phenomena are groups of experiences relating in the ways that experiences typically do relate to one another when they occur in different people perceiving the same thing. To make this intuitive notion precise, I define a "thinglike group" of experiences as a collection of experiences that boost each other's probability, and occur at the same location in time and space.

A main task of Chapter 4 is to develop a notion of an ideal spacetime in which the experiences of different conscious subjects can occur. To that end, I review the basic resources required to define physical spacetime, and define ideal spacetime in terms of parallel experiential resources. I then define ideal events, processes, objects, and worlds in terms of ideal spacetime and other phenomenal and topic-neutral concepts.[8]

The physical world, in my view, is the possibility for a certain ideal world. In **Chapter 5**, I explain what I mean by "possibility" in this context. Basically, the ideal world for which there's a possibility (in the relevant sense) is the one that our own actual experiences would belong to, if they belonged to an ideal world. To elaborate on this answer, I introduce the notion of phenomenal probabilities—objective conditional probabilities concerning purely experiential states of affairs—and explain how it allows us

[7] Those impatient to get to the main argument of the book may want to skip Ch. 2 on a first reading. However, reading this chapter will make it easier to navigate subsequent chapters, which are partly informed by a need to avoid objections to Mill's theory raised in Ch. 2.

[8] Ch. 4 includes a fairly high-level overview of how physicists define spacetime relations. An Appendix provides further details on scientific practice in this regard.

to avoid the objections that drove people away from phenomenalism in the mid-20th century.

At the end of Chapter 5, I officially state the phenomenalist theory I favor: the physical world we inhabit consists of the phenomenal probabilities in virtue of which a certain ideal world is the one that our own experiences have the greatest probability of belonging to.

In **Chapter 6**, I give a second argument for phenomenalism. There is an exceptionless correlation between physical things and possibilities for thinglike groups of experiences; the best explanation for this correlation is that physical things just *are* possibilities for thinglike groups of experiences; so, physical things are such possibilities.

This argument is similar to an influential argument for materialism: there's an exceptionless correlation between conscious experiences and certain types of brain states; the best explanation for this correlation is that conscious experiences just *are* such brain states; therefore, conscious experiences are brain states. The materialist argument invites various well-known objections, and the parallel argument for phenomenalism invites parallel objections. I argue that phenomenalists have responses to these objections that are unavailable to their materialist counterparts.[9]

In **Chapter 7**, I argue that phenomenalism sheds much-needed light on the relationship between the scientific world-view and our everyday experience of the world: the relationship is analogous to the one by which the deeper layers of a hypertext relate to the higher-level layers that link to them. I argue further that phenomenalism is an implication of the best explanation of how scientific vocabulary, especially the vocabulary of physics, gets its meaning.

Chapter 7 is also where I compare phenomenalism to structuralism, an alternative reduction of the physical deeply informed by modern physics. I argue that while the two theories have much in common, phenomenalism is preferable, since unlike structuralism it gives a satisfying account of the relationship between experience and the physical world.

Chapter 8 offers a phenomenalist perspective on the problem of consciousness, understood as reconciling the prima facie scarcity of consciousness with its prima facie fundamentality. We normally expect fundamental phenomena to play a foundational role in our ontology, serving as a reduction base for higher-level phenomena, but it's hard to see how a rare and

[9] The materialist argument originates with U.T. Place and J.J.C. Smart: see Place (1956) and Smart (1959).

highly localized phenomenon such as consciousness appears to be can play such a role. Consciousness seems fundamental, yet it can't plausibly play the foundational role we normally expect of fundamental phenomena. There seems to be some mystery here.

Materialists try to solve the mystery by denying that consciousness is fundamental; idealists and panpsychists try to solve it by denying that consciousness is scarce; dualists say that the mystery is unsolvable. I argue that phenomenalism not only is compatible with all of these options, but combines with most of them to yield a more satisfying overall view of the relationship between mind and body than we can obtain on any other terms.

Phenomenalism comes with a simple theory of perception: to perceive something is to have one of the experiences for which the thing is a possibility. In **Chapter 9**, I explain how this theory distinguishes veridical perceptual experience from non-veridical experience, as well as how it accounts for the possibility of phenomenally distinct perceptions of the same thing, and for the difference between perceiving something and knowing about it in a non-perceptual way. I argue that the phenomenalist theory of perception is an attractive alternative to the representational and naive realist theories that currently dominate the philosophy of perception.

Chapter 10 concludes the book with a summary of my overall case for phenomenalism, and a comparison between phenomenalism and David Lewis's Humean supervenience, a prominent theory whose main sources of anxiety—chance and experience—are the phenomenalist's metaphysical foundations.

2
Mill's Metaphysics

This chapter presents Mill's original theory in a way that brings out its advantages and its shortcomings, so that we'll know what to keep and what to change when working on an improved version of the theory in later chapters.

In §2.1, I trace the roots of Mill's phenomenalism back to traditional idealism, which Mill explicitly acknowledges as a major influence.[1]

Mill presents his phenomenalism as a simplification of the broadly Kantian metaphysics that dominated the metaphysical landscape of his day. This seems to me like an accurate representation, and in §2.2, I give an interpretation of Mill's theory that does justice to it.

Mill's phenomenalism had a significant following in late-nineteenth to early-twentieth-century metaphysics, and philosophers took it seriously into the 1960s. Since then, phenomenalism hasn't received much attention, and the attention it has received has tended to obscure rather than clarify Mill's position. In §2.3, I criticize several interpretations of Mill that cast his phenomenalism in an unfairly poor light.[2] This will give me an opportunity to dispel some common misconceptions about phenomenalism.

Though free from most of the flaws commonly attributed to it, Mill's theory does have two important shortcomings: it doesn't include very many details about the experiential phenomena that Mill says physical things are

[1] The etymology of "phenomenalism" is obscure; Grote (1865) is perhaps the earliest occurrence. (Sometimes people use "phenomenalism" to mean the view that there's nothing but experience; obviously, this isn't how I'm using the term: possibilities for experience exist, in the phenomenalist view, but they're not experiences.) Besides Mill, the most important phenomenalists are A.J. Ayer and C.I. Lewis: see Ayer (1936: 138–46; 1946) and Lewis (1946). H.H. Price defends a theory that has phenomenalist elements in Price (1932; 1940: 141–92).

[2] Geoffrey Scarre describes Mill's phenomenalism as "a confused mixture of idealist and immaterialist elements" (Scarre 1980: 194); according to John Skorupski, "if it were possible to make sense of [Mill's metaphysics] at all, it would have to be along the lines of transcendental idealism" (Skorupski 1989: 243); Andy Hamilton calls Mill's phenomenalism "an intriguing, elusive and puzzling doctrine" that is "transitional between eighteenth-century psychological explanation and twentieth-century semantic analysis" (Hamilton 1998: 172, 156); according to Nicholas Capaldi, "the only coherent line of development open to [Mill] is the movement toward Hegelian absolute idealism" (Capaldi 2004: 314); and Christopher Macleod suggests that Mill "never worked through the internal pressures of his own position with sufficient rigour to feel the push" in it towards Kantian or post-Kantian idealism (Macleod 2016: 33).

Phenomenalism: A Metaphysics of Chance and Experience. Michael Pelczar, Oxford University Press.
 DOI: 10.1093/oso/9780192868732.003.0002

possibilities for, and it's not very explicit about what it is for there to be possibilities for those phenomena. In §2.4, I discuss these shortcomings of Mill's theory and give a brief preview of how I plan to remedy them.

2.1 The genealogy of phenomenalism

A group of Bedouin travelers are camped around the Great Sphinx of Giza. One sees the Sphinx from due East, looking it straight in the face, another from the South, with a view of the Sphinx's right flank, another from the West, viewing the Sphinx from behind, etc. How do the travelers' visual experiences relate to each other?

One aspect of the relationship is qualitative. The traveler viewing the Sphinx from the East has an experience with a certain phenomenal shape and size, and the one viewing the Sphinx from the South has an experience with a different phenomenal shape and size, but there is some overlap in the spatial quality of the Eastern and Southern experiences, corresponding to the part of the Sphinx both viewers can see (the monument's southeastern quadrant). There's a similar overlap between the experience of the person who views the Sphinx from the South and the one who views it from the West; and likewise for the Southern viewer and a viewer standing between him and the Western viewer, etc. In short, the travelers' experiences fit together, like overlapping regions of the surface of the Sphinx itself.[3]

Another aspect of the relationship is modal. It's no accident that the travelers' experiences fit together the way they do. (So, it's different from a case where two people living far apart in time or space just happen to have phenomenally indistinguishable dreams of a Sphinx.) We might think of this in terms of a counterfactual or probabilistic interdependence among the travelers' experiences. For example, if the Southern Bedouin's experience had been as of a Sphinx half buried in sand, the Eastern Bedouin's experience would also have been as of a half-buried Sphinx, or at least the probability of the Eastern Bedouin's experience being as of a half-buried Sphinx given that the Southern Bedouin's experience is as of a half-buried Sphinx is high (or anyway higher than the unconditional probability of the Eastern Bedouin's

[3] The experiences don't have shapes and sizes in the same sense as the Sphinx has a size and shape, but they have qualia that invite description in geometric terms. The experiences are phenomenally Sphinx-shaped, in the same sense as they're phenomenally beige: beige as a visual image of a sand dune is beige, rather than as the dune itself is beige.

experience being as of a half-buried Sphinx). Changes in one Bedouin's experiences are likewise linked to changes in the others': if they're watching the Sphinx as the sun sets, the changes in each traveler's experiences mirror changes in the others'—the experiences phenomenally darken in tandem—in a way that's again not by chance, but supports relevant counterfactual conditionals or conditional probabilities.

We can think of the Bedouins' experiences as standing in these modal and qualitative relations without assuming the existence of anything else (Sphinx, light, human bodies, etc.). Maybe we need to call on an independent external world in order to *explain* why the experiences stand in these relations of interdependence and qualitative coherence, but the existence of the experiences, and their modal and qualitative relations to one another, are logically independent of any such explanatory factor. At least, it's not hard to imagine a world consisting of nothing but these experiences standing in these relations.

Classical idealism equates physical reality with a totality of experiences that relate in the same way as the Bedouin travellers'.

Leibniz was the first to develop this idea in detail. According to him, the basic building blocks of the universe are minds, or "monads," as he calls them. Each monad has a sequence of experiences that unfolds in accordance with a set of instructions.[4] The monads' mental lives have neither beginning nor end, so that associated with each monad there is an infinitely long stream of consciousness.[5] The monads don't affect one another, but they've been programmed by God so that their mental lives are coordinated, in the sense that their experiences are such as you'd expect of a population of conscious beings who perceive the same world (like the ideal observers described in the previous chapter). Physical reality just is this coordinated or harmonious unfolding of the monads' conscious mental lives:[6]

[4] In effect, a computer program: Leibniz describes monads as "very exact immaterial automata" (Leibniz 1998: 206).

[5] Leibniz (1714: 207; 1714/1989: 213).

[6] Here I'm setting aside Leibniz's late and obscure comments about "substantial bonds," which he makes when discussing the Catholic doctrine of transubstantiation in his correspondence with the Jesuit theologian Bartholomew Des Bosses: see e.g. Leibniz (1712/2007a: 225–7; 1712/2007e: 241–43; 1712/2007b; 1712/2007d: 294–99; 1712/2007c: 351). The theory I'm attributing to Leibniz is intended to capture the latest version that he worked out in detail, independent of concerns to reconcile his metaphysics with Catholic doctrine. For more on Leibniz on substantial bonds (*vincula substantialia*), see Jolley (2005: 81–3).

Philarète: There are even good grounds for doubting whether God made anything other than monads or substances without extension, and whether bodies are anything other than the phenomena [i.e., experiences] resulting from these substances. My friend [Leibniz], whose opinion I have just related, gives enough evidence that he leans in this direction, since he reduces everything to monads, or to simple substances and their modifications [experiential states], along with the phenomena that result from them, phenomena whose reality is indicated by their interconnections, something that distinguishes them from dreams.

(Leibniz, 1715/1989: 265)

According to Leibniz, the Bedouins' experiences are only parts of the Sphinx: the whole Sphinx includes not only their experiences, but many more (infinitely many, according to Leibniz). It includes experiences of the Sphinx viewed from every possible distance and angle, experiences of the Sphinx's underside, experiences of the Sphinx's solid interior, experiences of individual molecules of rock, etc.—intuitively, whatever experiences might characterize the conscious mental lives of Sphinx-observers with unlimited perceptual powers.

According to Leibniz, for there to be a physical world just is for there to be minds whose experiences fit together and unfold in tandem like the Bedouin travelers'. Physical reality isn't something over and above the monads and their experiences, but the monadic mental lives themselves, unfolding in this coherent and coordinated way:

Moreover, matter and motion are not so much substances or things as they are the phenomena of perceivers, the reality of which is located in the harmony of perceivers with themselves (at different times) and with other perceivers. (Leibniz 1704/2013: 307)

Leibniz understands the "harmony" of the monads in computational and geometric terms. Each monad's experiences counterfactually depend on each other's, in such a way that at each moment, the monads' experiences are as of the same spatial state of affairs viewed from different standpoints:

Finally, you ask, "Why are these appearances produced in me or any other true substance?" I say that subsequent appearances are produced from those preceding them in accordance with metaphysical and mathematical laws of eternal truth. But the reason why there are any such appearances at

> all is the same as the reason for the existence of the universe. For you can easily see that simple substances can be nothing other than the sources, i.e., the principles, and at the same time the subjects, of just as many series of perceptions unfolding themselves in order, expressing the same universe of phenomena with the greatest and most orderly variety.
>
> (Leibniz 1705/2013: 325)
>
> Just as the same city viewed from different directions appears entirely different and, as it were, multiplied perspectively, in just the same way it happens that, because of the infinite multitude of simple substances [monads], there are, as it were, just as many different universes, which are, nevertheless, only perspectives on a single one, corresponding to the different points of view of each monad. (Leibniz 1714/1989: 220)[7]

Leibniz makes it clear that he intends his theory as a reductive analysis of the physical:

> I do not really do away with body, but reduce it to what it is.
>
> (Leibniz 1705/2013: 319)

He also makes it clear that his view is *not* that there are two worlds, one physical and the other mental, that God harmonized *ab initio:* the pre-established harmony is a harmony among minds, and the physical world just is those minds with their thus-harmonized experiences:

> It is true that the things that happen in the soul must agree with those that happen outside the soul; but for this it is sufficient that those things that happen in one soul correspond both among themselves and with those things that happen in any other soul; and there is no need to posit something outside of all souls or monads. (Leibniz 2007: 257)

To conclude this summary of Leibniz's metaphysics: the truths about physical reality are truths about monads with various experiences standing in

[7] See also Leibniz (1712/1989: 199) "And the distinction between the appearance bodies have with respect to us and with respect to God is, in a certain way, like that between a drawing in perspective and a ground plan. For there are different drawings in perspective, depending upon the position of the viewer, while a ground plan or geometrical representation is unique. Indeed, God sees things exactly as they are in accordance with geometrical truth, although he also knows how everything appears to everything else, and so he eminently contains in himself all other appearances."

relations of geometric coherence and counterfactual interdependence, where this whole arrangement is due to God.

Berkeley has a similar view. According to Berkeley, a physical thing is a combination of experiences:

> Thus, for example, a certain colour, taste, smell, figure and consistence having been observed to go together, are accounted one distinct thing, signified by the name apple; other collections of ideas constitute a stone, a tree, a book, and the like sensible things. (Berkeley 1710/1901: §1)
>
> [A]ll sensible qualities are alike *sensations*, and... the objects of sense are nothing but those sensations, combined, blended, or (if one may so speak) concreted together. (Berkeley 1710/1901: §99)

Like Leibniz, Berkeley holds that experiences that constitute a physical thing differ from experiences that don't in virtue of their coherent and divinely coordinated relations:

> The ideas of Sense are more strong, lively, and distinct than those of the Imagination; they have likewise a steadiness, order, and coherence, and are not excited at random, as those which are the effects of human wills often are, but in a regular train or series—the admirable connexion whereof sufficiently testifies the wisdom and benevolence of its Author.
> (Berkeley 1710/1901: §30)

One way Berkeley differs from Leibniz is by not requiring physical things to include experiences belonging to different minds: according to Berkeley, most physical things—and all that are unobserved by created beings—consist entirely of experiences that occur in the mind of God.[8] Another is that Berkeley holds that a created mind perceives only a limited part of the physical universe, whereas Leibniz holds that every created mind perceives the entire physical universe (even though most of a created mind's experiences are inaccessible to clear and distinct introspection).[9]

[8] At least, this is Berkeley's view on one common interpretation: see Berkeley (1713/19: 424–5; 1710/1901: §§3, 6, 48, 90), and Winkler (1989: 204–37). According to an alternative interpretation, Berkeley doesn't think that God literally perceives physical things, but that He apprehends them in some non-perceptual way; see Mabbott (1931). I'm not going to weigh in on this interpretive question here.

[9] See Leibniz (1714/1989: 220–21): "Monads all go confusedly to infinity, to the whole; but they are limited and differentiated by the degrees of their distinct perceptions."

Despite these differences, Leibniz and Berkeley give us the same basic picture: physical things are combinations of experiences hanging together in a certain way, and the reason why there are such combinations is that God makes it so. Call this *traditional idealism*.[10]

Traditional idealism has three prima facie objectionable features.

First, it entails that if there were no conscious experience, there would be no physical things. If, as traditional idealists maintain, physical things are just suitable combinations of experiences, a world devoid of experiences can hardly contain any physical things. Yet, *pace* Berkeley, it's easy to conceive of situations in which many or even all of the physical things in our world exist, but there's no conscious experience. (One example of such a situation is a possible world just like ours except that it ends before any sentient life evolves; another is a "zombie world" just like ours physically but containing no consciousness.)

Second, traditional idealism entails that there is far more experience than we have evidence for. This implication is perhaps more severe for Leibniz than Berkeley, since in Berkeley's system all the excess experience is in God. Still, we normally don't think of our world as containing conscious perceptual experiences of each and every physical entity that ever exists, and even theists don't typically think of God as knowing the world through sense experience.

Third, traditional idealism offers a highly speculative explanation of the coherence and interdependence of ordinary sense experience: God. This problem is more severe for Berkeley than Leibniz, since only Berkeley identifies (most) physical things with divine experiences: Leibniz introduces God only to explain why there is all the experience there is, rather than different experience, or none. Still, from a modern standpoint, assigning God any important metaphysical role is controversial.

One way to see Kant's metaphysics is as an effort to preserve what's appealing about traditional idealism while avoiding these undesirable consequences.

Here's the Kantian picture: there are entities with the power to give us experiences. The only thing we can know about these entities—the so-called noumena—is that they have such powers. The noumena have their experience-causing powers regardless of whether they actually exercise them, and regardless of whether there are any minds for them to exercise

[10] However this may fare as a summary of traditional idealism, it's at least an accurate summary of how Mill understood Berkeley: see Mill (1871/1978: 459–65).

them on. For a physical thing—say, a tree—to exist is for it to be the case that if the noumena did exercise all of their experience-causing powers, then the total of all noumenally-induced experiences would include treeish experiences that related to all other experiences in a suitably coherent way.[11]

In effect, Kant's view is that the truths about physical reality are a logical consequence of the fact that something—further nature unknown—has the power to bring about the kind of experiential state of affairs that Leibniz and Berkeley identify with physical reality.

The Kantian theory avoids the first problem with traditional idealism by identifying physical things with noumenal powers to cause suitable combinations of experiences, rather than with actual combinations of experiences. Since a power can exist without being exercised, the Kantian theory allows that there could be an experienceless world physically identical to ours: that's just a world where the noumena have all the same experience-causing powers as in ours, but where, unlike in our world, they don't exercise any of those powers. We can express this difference between Kantian metaphysics and traditional idealism by saying that while both are *mind-first* metaphysics, in the sense that the concept of experience plays a central role in their analyses of matter, only traditional idealism implies that physical things are *mind-dependent,* relying on minds or experience for their existence.[12]

Kantian metaphysics avoids the second problem with traditional idealism, since it requires, for each physical thing, only that there be noumena with suitable experience-causing powers, rather than actual experiences. So Kant doesn't need to posit more experience than we already have evidence for.

Kantian metaphysics avoids the third problem with traditional idealism by refusing to attribute any features to the ultimate sources of experience besides powers to cause various experiences. Unlike Leibniz, who identifies the sources of experience with divinely-programmed monads, and Berkeley, who identifies the source of experience with God, Kant refuses to speculate about the nature of the ultimate sources of experience, beyond saying that

[11] Here I'm following Rae Langton's interpretation of Kant: see Langton (1998). I'm not going to attempt to defend Langton's interpretation or weigh it against others, of which there are many. It, or something very much like it, was how Mill understood Kantian metaphysics: see Mill (1865/1979: 7–9, 27).

[12] Actually, there's disagreement among Kant scholars over whether Kant himself thinks that physical things have a mind-independent existence. According to some interpretations of Kant, he holds that time and space are mind-dependent, committing him to the mind-dependence of physical things. Whatever the correct interpretation of Kant may be, there is a broadly Kantian view in which physical things exist regardless of whether there are any minds, and this is what I mean by "Kantian metaphysics" or "the Kantian view."

they have various experience-causing powers; indeed, he thinks that such speculation is futile.[13]

2.2 Kant without noumena

Mill's phenomenalism is a second step away from traditional idealism in the same direction as Kant's first step. It's the step of suspending judgement about the existence of *anything* that explains the propensity for experience to occur as it does. Why must there be something in addition to experiences and possibilities for experiences, in order for there to be physical things? No reason, says Mill. The noumena are an extravagance, just like the Gods and monads they're intended to replace. The possibilities are all we need for a physical world:

> Matter, then, may be defined, a Permanent Possibility of Sensation.[14]

Mill introduces his theory in the middle of a 500-page critique of the philosophy of William Hamilton, a philosopher whose work, in hindsight, hardly merited this level of attention. Chapters 11–13 of *An Examination of Sir William Hamilton's Philosophy* and their appendices are the main sources for Mill's phenomenalism.[15]

Mill gives the central theory in chapter 11. However, the official goal of that chapter, as indicated by its title—"The Psychological Theory of the Belief in an External World"—is to provide an alternative to Hamilton's account of the origins of our belief that there is an external world. According to Hamilton, the belief is "intuitive," by which he means something like innate, or in any event not formed on the basis of experience. Against this, Mill argues that our belief in an external world does come from experience.

The dispute between Mill and Hamilton over the origins of our belief in an external world is an unimportant footnote to the history of philosophy. Far more interesting is the metaphysical theory that Mill proposes in the course

[13] See Kant (1781/1998: 350).

[14] Mill (1865/1979: 183). That Mill saw his view as a simplification of Kantian metaphysics comes across in a letter to Herbert Spencer: "From the first I have wished to keep the peace with those whose belief in a substratum is simply the belief in an Unknowable" (Mill 1865/1972: 1090).

[15] See Mill (1865/1979: 177–249). He floats the general idea behind phenomenalism in Mill (1843/1974: 58), where he alludes to the view that a body "is, in short, a set of sensations, or rather, of possibilities of sensation, joined together according to a fixed law."

of developing his alternative to Hamilton's account of our belief in an external world. This is the part of Mill's discussion that interests us here.

Mill sees the physical world as consisting of possibilities for experiences to occur the way they typically occur in us when we perceive physical things. These possibilities are the world's physical furniture: an apple is a possibility for experiences to occur as they typically do when people perceive an apple, and likewise for all other physical things. The totality of all such possibilities is the physical world:

> The conception I form of the world existing at any moment, comprises, along with the sensations I am feeling, a countless variety of possibilities of sensation: namely, the whole of those which past observation tells me that I could, under any supposable circumstances, experience at this moment, together with an indefinite and illimitable multitude of others which though I do not know that I could, yet it is possible that I might, experience in circumstances not known to me. These various possibilities are the important thing to me in the world. (Mill 1865/1979: 179–80)

So far, Mill's view agrees with Kant's. The disagreement comes with Mill's claim that there's no need to posit any ultimate cause, ground, explanation, or substratum for the possibilities of sensation that he identifies with physical things:

> [T]here is, as I have said, for every statement which can be made concerning material phænomena in terms of the Realistic [i.e., Kantian[16]] theory, an equivalent meaning in terms of Sensation and Possibilities of Sensation alone, and a meaning which would justify all the same processes of thought. In fact, almost all philosophers who have narrowly examined the subject, have decided that Substance need only be postulated as a support for phænomena [i.e., experiences[17]], or as a bond of connexion to hold a group or series of otherwise unconnected phænomena together: let us only, then, think away the support, and suppose the phænomena to remain, and to be held together in the same groups and series by some other agency, or without any agency but an internal law, and every consequence follows

[16] See the reference to "German metaphysicians" in Mill (1865/1979: 196).

[17] In Mill's day, as today, the word "phenomenon" (or "phænomenon") was ambiguous between a more general sense, in which it just means entity, and a narrower sense in which it means a specifically conscious or "phenomenal" entity.

> without Substance, for the sake of which Substance was assumed. The Hindoos thought that the earth required to be supported by an elephant; but the earth turned out quite capable of supporting itself, and "hanging self-balanced" on its own "centre." Descartes thought that a material medium filling the whole space between the earth and the sun, was required to enable them to act on one another; but it has been found sufficient to suppose an immaterial law of attraction, and the medium and its vortices dropped off as superfluities. (Mill 1865/1979: 198)[18]

At the heart of Mill's theory is the identification of physical things with a certain type of possibility for conscious experience. Mill calls such possibilities "permanent possibilities of sensation," "certified possibilities of sensation," "guaranteed possibilities of sensation," and "potentialities of sensation," distinguished from "mere vague possibilities, which experience gives no warrant for reckoning upon."[19] Despite its catchy alliterative quality, "permanent possibility of sensation" is the least apt of these labels, since, as others have pointed out, Mill doesn't mean to identify impermanent physical things—soap bubbles, electrical discharges, etc.—with permanent (as opposed to impermanent) possibilities of sensation.[20] As I said in Chapter 1, I'll use "possibility of sensation," "possibility for experience," and "phenomenal possibility" interchangeably to refer to the kind of thing Mill has in mind.

Mill identifies physical things with possibilities for a certain type of experiential entity, which, for ease of reference, I'll call a *thinglike group* of experiences. In Mill's view, physical things are possibilities for thinglike groups of experiences.

A thinglike group of experiences is a group whose constituent experiences relate to one another as experiences typically do relate when they occur in various people perceiving the same physical thing. This is how the Bedouin travelers' experiences relate to one another in the example described earlier, and how the experiences of different ideal observers of some object or event relate to one another. Here's what Mill has to say about such groups:

[18] An even more striking example of dispensing with a "Substance" once regarded as indispensable was Gauss's proof that we can make sense of a curved space without thinking of it as embedded in a higher-dimension space (Gauss's celebrated *Theorema Egregium*: see Gauss 1827/1902). That Gauss and Mill hit upon this type of simplification at around the same time is suggestive, though I have found no evidence that Mill was aware of Gauss's discovery.

[19] See Mill (1865/1979: 179–84; 1871/1978: 459, 464).

[20] Skorupski (1989: 231); Hamilton (1998: 148).

> There is another important peculiarity of these certified or guaranteed possibilities of sensation: namely, that they have reference, not to single sensations, but to sensations joined together in groups. When we think of anything as a material substance, or body, we either have had, or we think that on some given supposition we should have, not some *one* sensation, but a great and even an indefinite number and variety of sensations, generally belonging to different senses, but so linked together, that the presence of one announces the possible presence at the very same instant of any or all of the rest. (Mill 1865/1979: 180)[21]

> Other people do not have our sensations exactly when and as we have them: but they have our possibilities of sensation; whatever indicates a present possibility of sensations to ourselves, indicates a present possibility of similar sensations to them, except so far as their organs of sensation may vary from the type of ours. (Mill 1865/1979: 182)

These remarks are suggestive, but they don't tell us all that much about thinglike groups of experiences. We're told that they're sensations "joined together in groups," but joined how? Talk of one experience "announcing the possible presence" of other experiences doesn't do much to answer this question. Answering it satisfactorily is an important item of unfinished business in Mill's metaphysics.

Turning to the second question: What does Mill mean by a "possibility of" or "potentiality for" some experience or group of experiences?

Although Mill never explicitly defines the relevant kind of possibility, his text suggests an intimate connection between possibilities of sensation and counterfactual conditionals of the form:

> *If such-and-such experiential state of affairs held, then such-and-such other experiential state of affairs would hold.*

[21] See also Mill (1865/1979: 201): "I have postulated, first, Sensations; secondly, succession and simultaneousness of sensations; thirdly, an uniform order in their succession and simultaneousness, such that they are united in groups, the component sensations of which are in such a relation to one another, that when we experience one, we are authorized to expect all the rest, conditionally on certain antecedent sensations called organic, belonging to the *kind* of each [e.g., muscular sensations of looking in a certain direction, or reaching towards something]. This is all we need postulate with regard to the groups, considered in themselves, or considered in relation to the perceiving Subject."

—where an experiential state of affairs is one that's fully describable in experiential and topic-neutral terms. Following Skorupski, I'll call such conditionals *sensation conditionals*.[22]

For a possibility of sensation to exist is for some sensation conditional or conjunction of sensation conditionals to be true. For example, Mill writes:

> I see a piece of white paper on a table. I go into another room. If the phænomenon always followed me, or if, when it did not follow me, I believed it to disappear *è rerum naturâ,* I should not believe it to be an external object. I should consider it as a phantom—a mere affection of my senses: I should not believe that there had been any Body there. But, though I have ceased to see it, I am persuaded that the paper is still there. I no longer have the sensations which it gave me; but I believe that when I again place myself in the circumstances in which I had those sensations, that is, when I go again into the room, I shall again have them; and further, that there has been no intervening moment at which this would not have been the case.
> (Mill 1865/1979: 179)

> I believe that Calcutta exists, though I do not perceive it, and that it would still exist if every percipient inhabitant were suddenly to leave the place, or be struck dead. But when I analyse the belief, all I find in it is, that were these events to take place, the Permanent Possibility of Sensation which I call Calcutta would still remain; that if I were suddenly transported to the banks of the Hoogly, I should still have the sensations which, if now present, would lead me to affirm that Calcutta exists here and now.
> (Mill 1865/1979: 184)

The counterfactual conditionals Mill uses in these passages don't transparently count as sensation conditionals, since their antecedents—"if I were to enter the room," "if I were transported to the banks of the Hoogly"—refer to physical objects and events. However, as Mill makes clear in response to his critics, he is merely trying to be reader-friendly in these passages. He is well aware that to get fully explicit sensation conditionals, he'd have to replace references to the room and the Hoogly with statements couched in purely experiential terms. He uses the words "the room" and "the Hoogly" only to direct his readers' attention to the experiences he has in mind. He assumes

[22] See Skorupski (1989: 231; 1994: 114). Topic-neutral terms express properties and relations that are restricted neither to the physical nor to the experiential domain; examples include the property of having parts, the property of resembling, and the relation of identity.

that his readers know roughly what kinds of experiences one would have if standing on the banks of the Hoogly; by "if I were transported to the banks of the Hoogly," Mill just means: if I were to have *those* sort of experiences.

The point seems simple enough, but it was evidently one that Mill's contemporaries had a hard time grasping. In the appendix to chapters 11 and 12 of the *Examination*, added after the first edition had been published and reviewed, Mill writes that his opponents

> one and all, think their case made out, if I employ, in any part of the exposition, the language of common life... If I say, that after we have seen a piece of paper on a table, our belief that it is still there during our absence means a belief that if we went again into the room we should see it, they cry out, Here is belief in Matter already assumed; the idea of going into a room implies belief in matter. [...] They forget that to go into a room [and similar terms] are expressions which have a meaning in the Psychological Theory[23] as well as in theirs; that every assertion that can be made about the external world, which means anything on the Realistic theory, has a parallel meaning on the Psychological. Going into a room, on the Psychological theory, is a mere series of sensations felt, and possibilities of sensation inferred, but distinguishable from every other combination of sensations and possibilities, and which, with others like to itself, forms as vast and variegated a picture of the universe as can be had on the other theory; indeed, as I maintain, the very same picture. The Psychological theory requires that we should have a conception of this series of actual and contingent sensations, as distinct from any other; but it does not require that we should have referred these sensations to a substance ulterior to all sensation or possibility of sensation. (Mill 1865/1979: 197)

I can tell you what I mean by "fool's gold" by saying that fool's gold is a worthless mineral that resembles gold, without implying that the existence of fool's gold presupposes the existence of gold; likewise, Mill can tell us what he means by "experiences as of being transported to the banks of the Hoogly" by saying that they're the sort of experiences you'd have if transported to the banks of the Hoogly, without implying that the existence of such experiences presupposes the existence of the Hoogly.[24]

[23] Mill often refers to phenomenalism as "the Psychological theory."

[24] This is related to Kripke's distinction between fixing the reference of a term and giving the meaning of a term: see Kripke (1980). See also Skorupski (1989: 235).

Call the facts expressed by true counterfactual conditionals "counterfacts."[25] One thing that distinguishes our world from other possible worlds is the counterfacts that hold in our world. (In this respect, counterfacts differ from facts about what's logically possible, which are the same in all possible worlds.)

Not all the facts of our world are counterfacts. That I'm having the experiences I am right now isn't a counterfact: it's an outright categorical fact. Likewise for all other facts about which experiences actually occur.

Among the counterfacts that characterize our world are counterfacts about experience, such as those expressed by true sensation conditionals. At least, our experiences exhibit patterns that make it reasonable for us to posit various counterfacts about experience.

So, we're given various outright facts about experience, including facts about how experience unfolds in our own conscious mental lives, and these facts warrant us in positing various counterfacts about experience, expressible by various sensation conditionals.

What other facts do we need to mention, in order to give a complete inventory of the facts of our world? According to Mill: none. In Mill's view, the facts of our world are exhausted by various outright facts about experience and various counterfacts about experience; the physical facts in particular are counterfacts about experience. For a physical thing to exist is just for certain counterfacts about experience to hold; for things to be as they are physically just is for certain sensation conditionals to be true. Whenever the truth of a sensation conditional calls for explanation, we can explain it by reference to the truth of further sensation conditionals.

To appreciate Mill's position, it's important to understand that a possibility of the relevant kind can actually exist, even if that for which it is a possibility doesn't actually exist. (In terms of sensation conditionals: the state of affairs that a sensation conditional describes can actually exist, even if the experiences that the conditional alludes to don't actually exist.) A possibility for breakage can exist even if nothing actually breaks; a possibility for atomic decay can exist even if no atom actually decays; a possibility for profit can exist even if no profit materializes.[26]

The word "possible" often gets used in contrast to "actual." That's not how "possible" is to be understood in the present context. The possibilities for experience that characterize our world are actual: they actually exist. This,

[25] Following McDermott (1999).

[26] H.H. Price makes this point in Price (1932: 275, fn. 1).

despite the fact that most of the experiences for which they are possibilities do not actually exist.

Define a *realized* possibility for experience as a possibility such that at least one of the experiences for which it is a possibility actually exists. If all the experiences for which it is a possibility actually exist, the possibility is fully realized. If none of the experiences for which a given possibility is a possibility actually exist, the possibility is fully unrealized. Most of our world's phenomenal possibilities are fully unrealized, according to Mill, but they're actual possibilities for all that.

Mill's identification of physical things with possibilities of sensation offers all the advantages of Kant's metaphysics, in a simpler package.

Like Kant's transcendental idealism, Mill's phenomenalism is a mind-first theory that does not entail that physical things are mind-dependent: for Mill, a world with all our world's physical features but no conscious experience is just a world where there exist all the possibilities of sensation that exist in our world, but in which the possibilities are all fully unrealized.[27]

Also like Kant, there's no pressure on Mill to posit more experience than we already have reason to believe exists. Just as Kantian noumena can exist without exercising their experience-causing powers, Millian possibilities for experience can exist without being realized.

Finally, unlike Leibniz and Berkeley, who identify God (or divinely-programmed monads) as the ultimate source of experience, Mill refuses to speculate about the ultimate source of experience, beyond holding that the regularities in our experience reflect the truth of various sensation conditionals. In this regard, Mill actually out-Kants Kant: where Kant posits noumena with experience-causing powers to account for the regularities in our experience, Mill posits only the truth of the sensation conditionals that Kant introduces the noumena to make true.

At this point I've presented what I take to be the central claim of Mill's phenomenalism, which is that the physical world is a possibility for things to be as traditional idealists say they actually are, where this possibility neither requires nor, as far as we know, has any ultimate ground, source, or explanation. A more aggressive version of phenomenalism would positively assert that the possibility has no ground, source, or explanation. Mill stops

[27] "The dormancy of all the possibilities [that constitute a physical body], while, as real possibilities guaranteeing one another, they continue to exist, constitutes, on the Psychological theory, the fact which is at the bottom of the assertion that the body is in existence when we are not perceiving it" (Mill 1865/1979: 202).

short of that, merely suspending judgement on whether any such ground, source, or explanation exists. But even if it does, his position is that the physical world consists of various possibilities for experience, not whatever might ground or explain the possibilities. If something explains why there exist the possibilities of sensation that do actually exist, that "something" *explains why* there is a physical world, but it isn't *identical with* the physical world. The physical world is identical with the possibilities.

I've said that Mill suspends judgement on whether our world's phenomenal possibilities have any *ultimate* explanation. By this, I mean that he suspends judgement on whether possibilities for experience *as a class* have any explanation; i.e., on whether anything explains why there are any such possibilities at all. This doesn't commit Mill to denying that possibilities for experience have explanations. In fact, Mill is committed to saying that many possibilities for experience do have explanations. Since Mill identifies all physical states of affairs with possibilities for thinglike groups of experiences, and since many (perhaps all) physical states of affairs have explanations in terms of other physical states of affairs, Mill has to say that many (perhaps all) possibilities for thinglike groups of experience have explanations *in terms of other such possibilities*.

This is a commitment that Mill readily accepts. According to him one possibility for a thinglike group of experiences can bring about another, and a possibility for a thinglike group of experiences typically is an effect of other such possibilities:

> The log which I saw on the fire an hour ago, has been consumed and has disappeared when I look again; the Possibilities of Sensation which I called by that name, are possibilities no longer. The ice which I placed in front of the fire at the same time, is now water; such Possibilities of Sensation as form part of the groups called ice and not of the groups called water, have ceased and given place to others. All this is intelligible without supposing the wood, the ice, or the water, to be anything underneath or beyond Permanent Possibilities of Sensation. Why, then, when I ascribe the disappearance of the wood, and the conversion of the ice into water, to the presence of the fire, must I suppose the fire to be something underneath a Possibility of Sensation? My experience informs me that those other Possibilities of Sensation do not vanish or change in the manner mentioned, unless another Possibility of Sensation known by the name of fire, has existed immediately before, and continued to exist simultaneously with the change. Changes in the Permanent Possibilities I find to have always

> for their antecedent conditions, other Permanent Possibilities, and to be connected with them by an order or law, as uniform as that which connects the elements of each group with another. (Mill 1865/1979: 202)

Take an ordinary physical thing, like a stalactite. The stalactite is the result of eons of mineral deposits that occur as water drips from the ceiling of a cavern. Mill recognizes that the stalactite is an effect of these chemical and mechanical processes, just as the puddle in front the fireplace is an effect of processes taking place in the fireplace and its environs. It's just that Mill sees both the stalactite and the processes that created it as possibilities of sensation.

The cavern ceiling and the dripping water are possibilities for certain experiences, the stalactite is a possibility for certain others, and the former possibilities bring about the latter. The possibilities for experience that are the physical causes bring about the possibility for experience that is the physical effect. In general, if there's a scientifically established causal explanation for a certain physical state of affairs, Mill can happily accept it. He'll just say that both the state of affairs and its physical causes are possibilities for experience.

Mill can afford to be neutral on whether *every* possibility for experience has a causal explanation. Maybe there are infinitely many physical events, each caused by other physical events, so that no event goes uncaused. Or maybe there are only finitely many physical events, in which case some physical events have no physical explanation. Likewise, there might be an infinite number of possibilities of sensation, each explained by other such possibilities, or it could be that there's only a finite number of possibilities of sensation, in which case Mill will say that some of the possibilities have no explanation (as far as we can know).

Mill also holds that some physical things—i.e., in his view, some possibilities of sensation—cause experiences. The story here is the same as for the stalactite, except that the effects are conscious experiences rather than (just) further possibilities of sensation:

> As already remarked, the constant antecedent of a sensation is seldom another sensation, or set of sensations, actually felt. It is much oftener the existence of a group of possibilities, not necessarily including any actual sensations, except such as are required to show that the possibilities are really present. Nor are actual sensations indispensable even for this purpose; for the presence of the object (which is nothing more than the

> immediate presence of the possibilities) may be made known to us by the very sensation which we refer to it as its effect. Thus, the real antecedent of an effect—the only antecedent which, being invariable and unconditional, we consider to be the cause—may be, not any sensation really felt, but solely the presence, at that or the immediately preceding moment, of a group of possibilities of sensation. (Mill 1865/1979: 186)

Here, Mill observes that a possibility of sensation can cause, or contribute to the causation of, one or more of the experiences for which it's a possibility. This is how it is in the Sphinx example, where the Bedouins' experiences are among those for which their cause (the Sphinx) is a possibility. In general, according to Mill, to perceive a thing just is to have one of the experiences for which the thing is a possibility. Experiences that aren't among those for which physical things are possibilities aren't veridical. This is how it is with non-perceptual experiences (or "purely mental feelings," as Mill calls them):

> We can, of course, think of all or any of our sensations [perceptual experiences] in relation to their Objects, that is, to the permanent groups of possibilities of sensation to which we mentally refer them. This is the main distinction between our sensations, and what we regard as our purely mental feelings. These we do not refer to any groups of Permanent Possibilities; and in regard to them the distinction of Subject and Object is merely nominal. These feelings have no Objects, except by metaphor. There is nothing but the feeling and its Subject. (Mill 1865/1979: 212)

Also included among the experiences that don't realize any of the possibilities that make up the physical world are non-veridical perceptual experiences. Distinguishing these from veridical perceptual experiences is critical, if Mill is to avoid counting the experiences we have in dreams and hallucinations as being among those for which real physical things are possibilities. Mill doesn't actually say much about this: his comments on it are limited to those quoted above on p. 24. When it comes to distinguishing veridical perceptual experience from non-veridical, Mill's view seems to be that Berkeley already showed how to distinguish them in a way consistent with idealism and phenomenalism, namely, in terms of how a given experience relates to all other experiences (or possible experiences):

> *Hylas:* But, according to your notions, what difference is there between real things, and chimeras formed by the imagination, or the visions of a dream—since they are all equally in the mind?

> *Philonous:* . . . [T]he visions of a dream, . . . though they should happen to be never so lively and natural, yet, by their not being connected, and of a piece with the preceding and subsequent transactions of our lives, they might easily be distinguished from realities. In short, by whatever method you distinguish *things* from *chimeras* on your scheme, the same, it is evident, will hold also upon mine. For, it must be, I presume, by some perceived difference; and I am not for depriving you of any one thing that you perceive.[28]

Berkeley doesn't tell us exactly how an experience must relate to other experiences to count as part of a physical thing, but he says that it's whatever relationship guides our actual judgements about which of our experiences are veridical and which aren't. That we do judge an experience veridical or not based on how it relates to other experiences follows, Berkeley argues, from the fact that such relations are all we ultimately have to go on in making such judgements.[29]

2.3 Mill misunderstood

I've offered a reading of Mill's phenomenalism as a reductive theory of the physical that simplifies the Kantian metaphysics prevalent at the time Mill wrote. Mill's theory, as I understand it, consists of the following claims:

(1) Physical things are possibilities for experience.

(2) As far as we know, nothing grounds, sustains, or otherwise underlies any of these possibilities, except as the case may be other possibilities for experience.

(3) The possibilities are for experiences that relate to one other as different people's experiences normally do when the people perceive the same thing, and that cohere with other possible experiences in the way your current experiences, but not the experiences you have in dreams, cohere with the rest of your experiences.

(4) For there to be a possibility for such a group is for a certain sensation conditional, or conjunction of sensation conditionals, to be true.

[28] Berkeley (1713/19: 452).

[29] I give a detailed phenomenalist account of the difference between veridical and non-veridical experience in Ch. 9.

(5) We perceive a physical thing by having one or more of the experiences for which the thing is a possibility.
(6) Some possibilities for experience cause other possibilities for experience, and some cause experiences.

(1) and (2) are the core of Mill's theory. (3) and (4) elucidate the notion of a permanent possibility of sensation central to the theory. (5) gives the theory's account of perception. (6) follows from (1), given the widely-accepted facts that many physical events have physical causes, and some physical events have experiential effects.

The phenomenalist theory I favor differs from Mill's mainly in how it fills out various details related to (3) and (4). To continue with our discussion of Mill, I want to consider three alternative readings of his metaphysics. These agree with my reading in broad outline, but differ from it on several important points, and their authors all take Mill's theory to have problems that it doesn't have, if I've understood Mill correctly. Considering these alternative interpretations of Mill will give me an opportunity to address some important misconceptions about the phenomenalist worldview.

2.3.1 Scarre's reading

The first reading of Mill I want to consider is Geoffrey Scarre's. Scarre discusses a number of possible reconstructions of Mill's phenomenalism, but the one he settles on as most likely to reflect Mill's intentions is much the same as the one I gave above: physical things are (mostly unrealized) possibilities for experience that have, as far as we know, no basis in anything, other than further possibilities for experience. However, Scarre claims to find serious flaws in Mill's theory that are absent from the theory I've attributed to Mill. Here, I'll discuss three of the main objections that Scarre raises against Mill.

2.3.1.1 An inadmissible switch of modalities

Scarre's first objection is that "Mill's theory apparently makes an inadmissible switch of modalities, from actual objects to possibilities of sensation."[30] He approvingly quotes G.J. Warnock:

[30] Scarre (1980: 184).

> it [seems] inadequate to regard the actual table in my empty study as a possibility of sensation. For surely, when I say there is a table in my study, I am saying what actually is the case, not mentioning merely the possibility of anything. (Warnock 1969: 227)[31]

According to Scarre, "Only if reification of permanent possibilities of sensation can be made plausible will this impression [of an inadmissible switch of modalities] be countered."[32]

These remarks suggest an imperfect grasp of Mill's position, as I understand it. Mill agrees that when Warnock says there's a table in his study, he's saying what is actually the case: namely, that there actually is a certain possibility of sensation. As pointed out earlier, there is no problem with the idea that there exist possibilities for non-existent things. According to Mill, an unperceived thing is a possibility for experiences, none of which actually exists. The *possibility* actually exists—it's a feature of the actual world—it's just that the experiences for which it is a possibility do not.[33]

It's strange to speak of "reifying" possibilities of sensation in this context. Once Mill has introduced the possibilities, there's no need for any further reification. To posit the possibilities just is to posit their reality. Anyway, Mill can hardly avoid reifying the possibilities, since he identifies them with physical things.

2.3.1.2 Non-mental laws

A second objection Scarre raises is that Mill can't consistently hold both that physical facts reduce to facts about possibilities for experience, and that possibilities for experience are subject to non-mental laws: "for if reality is mental in nature, its laws should be mental too."[34]

It's potentially misleading to say that Mill holds that reality is mental in nature. That makes Mill sound like Berkeley. Presumably Scarre means that Mill holds that physical reality consists of *possibilities* for something mental. Let's understand "mental" broadly to include not just experiences but possibilities for experience, which is how I take it Scarre is using the word here.

[31] Quoted in Scarre (1980: 184). In the same spirit, David Armstrong rejects phenomenalism on the grounds that "unfulfilled possibilities are nothings" (Armstrong 1961: 56).

[32] Scarre (1980: 184).

[33] Ayer makes substantially the same point (Ayer 1946: 180–81).

[34] Scarre (1980: 194).

Mill can't consistently say that possibilities of sensation are governed by laws whose existence presupposes the existence of irreducibly physical things. But Mill doesn't say that. The non-mental laws that govern possibilities of sensation, according to Mill, are laws of causation. These laws are non-mental only in the sense that they're topic-neutral, applying equally to entities in different domains. They're non-mental in the same sense as the laws of arithmetic. Saying that permanent possibilities of sensation are subject to arithmetical laws doesn't commit us to denying that the possibilities are mental; no more does saying that they're subject to causal laws commit us to denying this. Otherwise, Berkeley couldn't say that God causes minds to have experiences, and a materialist couldn't appeal to causal laws, since in whatever sense causality is non-mental, it's also non-physical.

2.3.1.3 Idealism vs. immaterialism

Scarre's main objection to Mill is that

> [Mill's] theory was not a consistent one, but rather a confused mixture of idealist and immaterialist elements. (Scarre 1980: 194)

By "immaterialism," Scarre means any theory that reduces material things to something broadly mental. By "idealism," he means any theory that combines immaterialism with a further mentalistic reduction of time and space. Given these definitions, Mill is clearly committed to immaterialism. It's also true, given these definitions, that he rejects idealism, since he's explicitly anti-reductive about time.[35] However, this says more about Scarre's definition of "idealism" than about Mill's position. Berkeley is also anti-reductive about time; at least, he offers no reduction of it, and no hint as to how such a reduction might go. It would seem to be a poor definition of "idealism" that doesn't let us apply that word to Berkeley's metaphysics.

In any event, Scarre suggests that Mill sometimes accepts, but sometimes rejects, a reductive view of space. That Mill sometimes accepts a reductive view of space is uncontroversial.[36] The question is whether he ever commits himself to denying such a view.

[35] Mill (1865/1979: 198–9). It's actually doubtful that Mill has to take an anti-reductive stance on time; elsewhere (Pelczar 2015: 56–83) I've argued that he does not.

[36] See, e.g., Mill (1865/1979: 221–2).

According to Scarre, he does, by claiming that permanent possibilities of sensation are public, in the sense that more than one person can perceive the same permanent possibility of sensation:

> [Mill's] denial of the objectivity of space greatly hampers a view which he definitely did wish to support, namely, that permanent possibilities of sensation are common to different subjects of experience. He often reiterated the claim that while actual sensations are private to individuals, possibilities of sensation have a public status. But it is extremely hard to accord any content to the notion that possibilities of sensation are common to different subjects if no reference is allowed to their situation in an objective, common space. (Scarre 1980: 191)
>
> Different subjects might experience qualitatively similar permanent possibilities in their own private spaces, but a public space is required for them to experience numerically the same ones. (Scarre 1980: 191)

It's not immediately clear that Mill faces a serious problem here. A possibility of sensation is public, provided that different people can have experiences for which it's a possibility. In Mill's view, different people perceive the same physical thing when they have some of the experiences that the thing is a possibility for. There's no immediately obvious need for Mill to appeal to an independently-given space for different subjects' experiences to occur in.

However, I think Scarre's criticism does point to a real problem for Mill.

The problem actually goes all the way back to the classical idealists. According to Berkeley, a physical thing is a combination of experiences: he speaks of a "complexion of ideas," and of sensations being "concreted together" to form a physical thing.[37] But how, in Berkeley's view, can different subjects' experiences be parts of the same object or event? How can Berkeley account for the publicity or intersubjectivity of physical things?[38]

Berkeley takes for granted a common time for all experiences, but simultaneity and qualitative coherence aren't sufficient to make a group

[37] Berkeley (1713/19: 459; 1710/1982: §99).

[38] Berkeley makes this question even harder for himself by denying that sight and touch can have a common object (1709/1901: §49). He even denies that it's possible for naked-eye observation to have the same object as microscopic observation (1713/19: 463–4). I'm going to overlook these oddities of Berkeley's position: as we'll see in a moment, he has trouble enough without them.

of sensations experiences of the same thing. There's a difference between two people seeing the same mountain at the same time, and two people seeing different but indistinguishable mountains at the same time.[39]

According to Mill, physical things are possibilities for sensations "joined together in groups" and "so linked together, that the presence of one announces the possible presence at the very same instant of any or all of the rest." Talk of one experience announcing the possible presence of other experiences is metaphorical, but it's clear enough what Mill means: a physical thing is a possibility for experiences to occur in the simultaneous and counterfactually interdependent way that people's experiences typically do occur when they perceive the same thing at the same time.

Mill's account faces essentially the same problem as Berkeley's. Suppose two people are watching the same television broadcast on different televisions. Their visual experiences occur simultaneously, and each person's experiences depend on the other's, in the same way they would if they were viewing the program on the same screen. (The experiences don't affect one another, but they stand in relations of counterfactual or probabilistic dependence.) If people perceived the same thing whenever their experiences were simultaneous and interdependent, we'd have to say that the people watching the program are perceiving the same screen images. But they're not.

What Mill (and Berkeley) could really use is an ideal spacetime capable of containing the experiences of different conscious subjects. Then they could say that people looking at distinct but indistinguishable mountains, or viewing the same television broadcast on different televisions, have experiences that occur at different locations in ideal space. I pursue this suggestion in Chapter 4, where I describe an intersubjective ideal spacetime modeled on the physical spacetime of modern physics.

[39] Lisa Downing suggests that "One way to dissolve this difficulty is to recall that objects are bundles of ideas. Although two people cannot perceive/have the numerically same idea, they can perceive the same object, assuming that perceiving a component of the bundle suffices for perception of the bundle" (2011: 27) But this doesn't dissolve the problem. It just relocates it. Now the question is: when do different people's experiences belong to the same bundle? (The answer can't be: when they belong to the same set of simultaneous and qualitatively coherent experiences. The experiences of two people perceiving distinct but indistinguishable mountains belong to such a set.) Leibniz solves the problem, or tries to, by appealing to the principle that it's impossible for two things to be exactly alike. However, it's doubtful that this principle (the identity of indiscernibles) is true.

2.3.2 Skorupski's reading

The next reading of Mill I want to consider is John Skorupski's. Skorupski's interpretation agrees with the one I've given, in broad outline, but unlike me, Skorupski thinks that Mill's phenomenalism conflicts with a naturalistic world-view in which minds and experiences arise from interactions between suitably constituted bodies and their physical environments. Specifically, Skorupski thinks that Mill's phenomenalism is incompatible with both of the following claims:

> *Psychological naturalism:*
>
> Minds and experiences have physical causes.
>
> *Scientific realism:*
>
> Our best scientific theories are true, or at least might be, including in what they say about features of the world that we can't observe.

Skorupski gives three arguments for the alleged conflict between Mill's phenomenalism and a naturalistic outlook. The first aims to show that phenomenalism conflicts with both psychological naturalism and scientific realism:

- A1 The relata of causal relations—causes and effects—must be categorical entities.[40]
- A2 Phenomenalism implies that physical things (objects, events, processes, etc.) are not categorical entities.
- A3 Therefore phenomenalism implies that physical things can't stand in causal relations.

—where categorical entities are those that aren't irreducibly modal, and modal entities are things like possibilities, probabilities, potentials, dispositions, propensities, and powers.

The conclusion of this argument implies a conflict between phenomenalism and psychological naturalism, as well as between phenomenalism and scientific realism (assuming that our best scientific theories work only if

[40] "It is a common-sense requirement of causation that it should relate *categorical* conditions, states, or events" (Skorupski 1989: 242–3); "[Scientific reasoning] enjoins us to look for a causal connection between underlying *categorical* states and processes to explain them" (p. 243, emphasis original).

physical events stand in causal relations). The argument's second premise is clearly true. What about the first premise? In support of it, Skorupski asks, rhetorically:

> How can Permanent Possibilities of Sensation cause anything?
>
> (Skorupski 1989: 243)

It seems fair to respond by asking how *anything* can cause anything. Whatever the relation of causality is, there are no obvious restrictions on what it might relate (apart from a restriction to non-abstract entities). As Hume says, "any thing may produce any thing" (at least if we're talking about non-abstract things).[41]

Far from being a necessary truth that causality relates only categorical relata, it's an open scientific question whether the relata of *actual* causal relations are categorical. According to one mainstream interpretation of quantum mechanics, the world's fundamental physical features are irreducibly probabilistic. If this is correct, causal relations among physical phenomena are, at bottom, causal relations among non-categorical phenomena (probability amplitudes or wave functions). Correct or not, it isn't something we can rule out a priori.

According to Mill, the relata of causal relationships among physical things are various possibilities for experience. As for the causal relation itself, Mill's phenomenalism is compatible with all major accounts of it, as far as I can tell, without being committed to any of them (though Mill himself seems to favor a Humean theory). Mill's position is consistent with the standard view of causation as a topic-neutral relation. So, the answer to Skorupski's question is: by standing to something in whatever relation the best theory of causation says the causal relation is.

Skorupski's second argument focuses specifically on scientific realism:

B1 If scientific realism is true, it's possible for scientific inquiry to reveal how things are at a level of physical reality that beings like us, with our human perceptual capacities, cannot perceive.[42]

B2 Phenomenalism implies that it's impossible for us to know how things are at a level we can't perceive.[43]

B3 Therefore, phenomenalism is incompatible with scientific realism.

The problem with this argument is B2.

[41] Hume (1739/2007: 1.3.15.1).

[42] Skorupski (1989: 203).

[43] Skorupski (1989: 205–6).

Mill's phenomenalism is compatible with there being physical things that we (ordinary human beings) can't perceive. Such things are possibilities for humanly impossible experiences. It's uncontroversial that some creatures have experiences that ordinary humans can't have, and there's no obvious limit to how many forms of non-human (or superhuman) experience are possible. So there's no obvious impediment to Mill identifying humanly imperceptible things with possibilities for humanly impossible experiences.[44]

B2 is false, because Mill's phenomenalism permits us to learn about humanly imperceptible things through their humanly perceptible effects. We discover how things are at a level we can't perceive by having experiences that realize phenomenal possibilities that are effects of possibilities for humanly impossible experiences.[45]

A phenomenalist could try to reduce facts about humanly imperceptible things to facts about possibilities for human experiences; for example, one might try to equate atoms with possibilities for the sort of experiences that prompt scientists to infer that there are atoms. However, it's unclear how this can work for an individual atom. Alternatively, a phenomenalist could take a scientific antirealist stance on humanly imperceptible things (you can hardly be an antirealist about individual atoms without being an antirealist about atoms en masse). But Mill doesn't endorse either scientific antirealism or an identification of humanly imperceptible things with possibilities for humanly possible experiences, and neither to my knowledge do any of the major phenomenalists who came after him.[46]

Skorupski's third argument for a conflict between scientific realism and Mill's phenomenalism is as follows:

C1 Phenomenalism is compatible with scientific realism only if it's compatible with the claim that beings with radically different perceptual

[44] See Mill (1865/1979: 182), where he speaks of sentient beings whose experiences differ from ours due to having different types of sense organs, and p. 192, where he characterizes a hypothetical Divine Mind as having "superhuman thoughts and feelings." He definitely doesn't think that the world's contents are limited to what we can imagine: see p. 68.

[45] This way of learning about humanly imperceptible things involves an inference to the best explanation. Mill himself (1865/1979: 78) suggests such an inference.

[46] Ayer, Lewis, and Price—the major phenomenalists of the 20th century—take no firm stance on the status of humanly imperceptible things, and this can give the impression that they think there are no such things, or that they would reduce their existence to the truth of counterfactuals about the sort of experiences that lead scientists to posit them. The truth, I suspect, is that giving a phenomenalist theory of humanly observable things was a big enough challenge that historic phenomenalists never advanced beyond it to consider the humanly unobservable case in detail.

modalities from ours could perceive and scientifically investigate the same world we do.[47]

C2 Phenomenalism is not compatible with the claim that beings with radically different perceptual modalities from ours could perceive and scientifically investigate the same world we do.[48]

C3 Therefore, phenomenalism isn't compatible with scientific realism.

The problem with this argument is C2. Mill *can* say that beings with radically different perceptual modalities from ours—phenomenal aliens—can perceive and investigate the same world that we perceive and investigate. This only requires him to add possible alien experiences to those for which physical things are possibilities.

Consider people whose experiences are much like ours, except that their visual phenomenology is color-inverted relative to ours. Intuitively, their experiences do as good a job as ours of informing them about the world's chromatic features. Our color-inverts' experiences relate to ours like otherwise identical maps with inverted color schemes, one showing land in yellow and water in blue, the other land in blue and water in yellow.

There's nothing wrong with perceiving ripe strawberries by having phenomenally green experiences, provided you also perceive ripe tomatoes that way, just as there's nothing wrong with navigating by a map that depicts the Atlantic Ocean in yellow, provided that it also depicts the English Channel in yellow. Different ways of mapping experiential properties and relations onto physical properties and relations are just that: different, without any of them being in any important sense better than any of the others.[49]

I've argued that Mill sees the physical world as a possibility for the sort of experiential state of affairs that traditional idealists identify with the physical world. The possibility of color-inverted experience merely requires us to take an expanded view of what that state of affairs includes. A tangerine is a possibility for a group of experiences that includes not just phenomenally orange experiences like ours, but phenomenally blue experiences like those of our chromatic inverts.

[47] (Skorupski 1989 203–204). [48] Skorupski (1989: 207).

[49] This is consistent with, but does not entail, the claim that normal color experience has the same representational content as inverted color experience. If, as some argue, a conscious experience always represents its own phenomenal features (in addition to whatever else it might represent), then, since the inverts' experiences have different phenomenal colors from ours, their visual experiences have different representational contents from ours. But that's consistent with their experiences representing the same external physical states of affairs. (For the view that conscious experiences represent themselves, see Kriegel 2009.)

This raises an important question: where do we draw the line between experiences of a tangerine that are veridical though phenomenally different from ours, and non-veridical experiences of a tangerine? If the tangerine looks blue to me due to a trick of the light or the effects of a drug, it seems that I experience the tangerine as having a color it doesn't really have, even though the phenomenal quality of my experience is the same as that of a chromatic invert's experience of the tangerine.

I give a detailed answer to this question in Chapter 9, where I discuss the phenomenalist theory of perception. The basic idea is that whenever an experience is inaccurate in some respect, there's a possibility for further experience that reveals the inaccuracy. For example, there's a possibility for my experience of the tangerine to change from phenomenally blue to phenomenally orange as the light returns to normal or the drug wears off, and a concurrent possibility for experience of the tangerine's microphysical surface features remaining the same throughout the changes in lighting or drug effects. The difference between my blue experience of the tangerine and the chromatic inverts' experiences of it is that there's no possibility of experience revealing the inverts' experiences as inaccurate with regard to color.[50]

As with people whose experiences are color-inverted relative to ours, so with people whose experiences differ from ours more radically, provided that the phenomenal qualities of their experiences map onto ours as the inverts' do. We can imagine beings who perceive physical colors by having auditory experiences, and physical sounds by having visual experiences; we can imagine beings who perceive colors by having experiences that we can't even imagine. If there's a possibility for such experiences in our world, Mill has the resources to incorporate them into his theory. In all such cases, he can say the aliens perceive and investigate the same world as us, by having experiences that partly realize the same (expanded) possibilities that our experiences partly realize. A strawberry is a possibility for a totality of interrelated experiences of the sort that would occur if a multitude of human and alien observers perceived the same strawberry. We perceive a strawberry by realizing some of the possibilities in this totality; the aliens perceive it by realizing others.[51]

[50] For further discussion, see pp. 175–178.

[51] In Ch. 7, I go into the relationship between human and possible alien experience in more detail: see pp. 141–143.

2.3.3 Hamilton's reading

The last reading of Mill's metaphysics I'll consider is Andy Hamilton's. On some points Hamilton agrees with Skorupski and Scarre. Like Skorupski, Hamilton thinks that Mill's phenomenalism is incompatible with his psychological naturalism; like Scarre, he thinks Mill is guilty of an illicit "reification" of possibilities for experience; like both Skorupski and Scarre, he thinks that physical things are categorical entities and so cannot be identified with bare possibilities.[52] Here I focus on aspects of Hamilton's reading that distinguish it from Skorupski's and Scarre's.

Hamilton makes three claims about Mill's metaphysics that I want to consider: that Mill adopts a stance of "ontological neutrality" towards various theories that prima facie conflict with his own; that Mill is committed to holding that the regularities we find in our experience are all "brute"; and that Mill's phenomenalism makes physical things objectionably mind-dependent.

2.3.3.1 Ontological neutrality

According to Hamilton, Mill is an "irenic philosopher, seeking to harmonize apparently rival metaphysical positions."[53] In this view, Mill favors "a novel and sophisticated form of reduction, if it is reduction at all—the ontological neutrality later advocated by Mach, Carnap and Schlick."[54]

> On the most developed statement of this view, the conflict between phenomenalism and other metaphysical positions, and that between these and common sense, are empty of content. As Carnap wrote: "the realistic and the [phenomenalist or physicalist] constructional languages have actually the same meaning. . . . [Once they are] recognized as nothing but two different languages which express the same state of affairs, several, perhaps even most, epistemological disputes become pointless."
>
> (Hamilton 1998: 154)

Hamilton offers two pieces of textual support for attributing an attitude of ontological neutrality to Mill.

The first is Mill's claim, in the passage quoted on p. 25 above, that the possibilities for experience that phenomenalism identifies with the physical

[52] See Hamilton (1998: 157–9). [53] Hamilton (1998: 154).
[54] Hamilton (1998: 154).

world form "as vast and variegated a picture of the universe as can be had on the other [Kantian] theory; indeed, as I maintain, the very same picture." Hamilton takes this to mean that there's no difference between phenomenalist and Kantian metaphysics. But all Mill is saying here is that phenomenalists and Kantians completely agree about the world's physical contents, disagreeing only about the correct metaphysical analysis of those contents. Mill immediately goes on to say that his theory "requires that we should have a conception of this series of actual and contingent sensations, as distinct from any other; but it does not require that we should have referred these sensations to a substance ulterior to all sensation or possibility of sensation." Since the Kantian theory *does* require that we refer the sensations to a substance ulterior to all sensation or possibility of sensation, Mill clearly thinks that his theory is an alternative to Kant's, not a mere restatement of Kant's theory in different language.

The other piece of evidence Hamilton offers in support of his claim that Mill advocates ontological neutrality is Mill's claim that phenomenalism and Kantian noumenalism have the same practical consequences:

> It is hardly necessary [. . .] to mention Dr. Johnson, or any one else who resorts to the *argumentum baculinum* of knocking a stick against the ground. Sir W. Hamilton, a far subtler thinker than any of these, never reasons in this manner. He never supposes that a disbeliever in what he means by Matter [i.e., a noumenal source of experience], ought in consistency to act in any different mode from those who believe in it. He knew that the belief on which all the practical consequences depend, is the belief in Permanent Possibilities of Sensation, and that if nobody believed in a material universe in any other sense, life would go on exactly as it now does. (Mill 1865/1979: 183–4)

It's hard to see how these remarks support an attribution of ontological neutrality to Mill. Once again, it's more natural to understand them as making the point that phenomenalists and Kantians (or Sir W. Hamiltonians) agree about the world's physical contents, a theory's implications in this regard being what determines its practical consequences. That Mill doesn't intend his remarks as a statement of ontological neutrality seems clear from the fact that he contrasts those who accept Sir W. Hamilton's theory of matter from those, like Mill, who disbelieve that theory. If Mill thinks the theories are equivalent, why does he disbelieve Hamilton's while believing his own?

2.3.3.2 Brute regularities in experience

A second criticism Hamilton raises against Mill's phenomenalism is that

> phenomenalists must regard regularities in sensations simply as "brute" facts. (Hamilton 1998: 159)

Mill can respond to this in either of two ways, depending on whether by "brute" Hamilton means irreducible, or inexplicable. If by "brute" Hamilton means irreducible, Mill agrees that regularities in sensations are brute facts: they're brute, in that the experiences that figure in the regularities are irreducible. This is just a corollary of Mill's anti-reductionism about consciousness. But if by "brute" Hamilton means inexplicable, Mill denies that regularities in sensations are brute facts.

Mill accepts that our experiences have neural causes: this is his psychological naturalism. Some brain states and processes (which, like any physical state or process, are permanent possibilities of sensation in Mill's view) cause conscious experiences. We can explain why we have the experiences we do, exhibiting whatever regularities they exhibit, by reference to the fact that we have brains in which such states and processes occur in certain patterns. This isn't a reductive explanation, but it is an explanation. (Likewise, we can explain present events in terms of past events without reducing facts about the present to facts about the past.)

Mill also has the resources to explain intersubjective regularities in experience. To adapt one of Mill's own examples, suppose a boy is playing frisbee with his dog. As the frisbee approaches the dog, there's an intersubjective regularity in experience: as the boy's images of the frisbee shrink (occupy less of his visual field), the dog's images of the frisbee correspondingly grow.[55]

Mill can explain this regularity as follows. A plastic disc is moving away from a boy toward a dog. Light reflected from the disc enters each animal's eyes, stimulating their retinal surfaces. The size of the stimulation area is proportional to the distance between the disc and each animal. Since this distance is inversely correlated (the closer the frisbee is to the boy, the farther it is from the dog), the amounts of retinal stimulation are also inversely correlated. Because of how brains are connected to eyes, the inversely-sized retinal stimulations cause the animals' retinotopic maps (located in their

[55] For Mill's example, see the conclusion of the long footnote ending on p. 207 of Mill (1865/1979).

visual cortices) to enter states that are inversely correlated in terms of how many of the maps' constituent neurons are activated. Since these retinotopic states are the neural correlates of the animals' visual experiences, the boy's visual images of the frisbee shrink as the dog's grow.

Mill can accept this explanation. He would only add that the brains, the eyes, and the frisbee are all permanent possibilities of sensation. The human and canine experiences are coordinated because there's an objective propensity for the boy's experiences to change in certain ways given that the dog's experiences change in certain ways (and vice versa), but this propensity isn't a brute fact: we can explain it as above by reference to the possibilities of sensation that constitute the boy's body, the dog's body, and various features of their physical environment.

2.3.3.3 Mind-dependence

On my reading of Mill, he doesn't think that physical reality is mind-dependent, since he identifies physical things with possibilities that could exist even if there were no minds. Against this, Hamilton suggests that Mill is committed to asserting the mind-dependence of physical things. On this reading of Mill,

> the possibilities [of sensation] are "certified beliefs" had by minds, and it is we who "reify" them, as the "psychological theory" shows. This interpretation is supported by Mill's assertion, in his review of Berkeley, that the possibilities "are not a positive entity... they did not exist as sensations, but as a guaranteed belief." (Hamilton 1998: 158)

If Hamilton is right, Mill's theory is deeply flawed. The main advantage of Mill's theory over Berkeley's, as Mill himself emphasizes in the essay Hamilton cites, is that phenomenalism, unlike Berkeleyan idealism, is compatible with the claim that our world's physical contents could exist even if there were no minds or experiences (human or divine). If Mill were to construe permanent possibilities of sensation as beliefs, he would completely throw away this advantage: if physical things are permanent possibilities of sensation, and permanent possibilities of sensation are beliefs, then physical things can't exist in the absence of beliefs. This is mind-dependence with a vengeance.

Hamilton's evidence for the claim that Mill construes permanent possibilities of sensation as beliefs is a passage from Mill's essay on Berkeley. Here is the passage in full:

> [Berkeley] supposed that the actual object of a sensible perception, though, on his own showing, only a group of sensations, and suspended so far as we are concerned when we cease to perceive it, comes back literally the same the next time it is perceived by us; and, being the same, must have been kept in existence in another mind. He did not see clearly that the sensations I have to-day are not the same as those I had yesterday, which are gone, never to return; but are only exactly similar; and that what has been kept in continuous existence is but a potentiality of having such sensations, or, to express it in other words, a law of uniformity in nature, by virtue of which similar sensations might and would have recurred, at any intermediate time, under similar conditions. These sensations, which I did not have, but which experience teaches me that I might have had at any time during the intermission of my actual sensations, are not a positive entity subsisting through that time: they did not exist as sensations, but as a guaranteed belief; implying constancy in the order of phenomena, but not a spiritual substance for the phenomena to dwell in when not present to my own mind. (Mill 1871/1978: 464)

This passage seems like a thin basis on which to saddle Mill with a commitment that would drag down his whole metaphysics. In any event, we can read the passage in a way that doesn't commit Mill to identifying physical things with beliefs.

Suppose I perceived some physical thing yesterday, and I perceive it again today, but nobody perceived it in the intervening time. According to Mill, there were no actual sensations corresponding to the thing during that intermediate time, but there were potentialities of sensations: the potentialities that constituted the physical thing during that time. When Mill says that the (logically) possible sensations for which these potentialities were possibilities "did not exist as sensations, but as a guaranteed belief," we can take him to mean that the only *psychological* reality that these logically possible sensations had during the intermediate time took the form of a reasonable belief of mine that possibilities for those experiences existed during that time.

Apart from the fact that this reading of the passage makes it consistent with the rest of Mill's metaphysics, it's plausible given the passage's textual setting. Mill is criticizing Berkeley for supposing there to be a numerical identity where in fact there is only a qualitative indistinguishability (between earlier and later sensations). When Mill attributes an error to another philosopher, he usually proposes an explanation for why that philosopher made the error. One possible explanation for Berkeley's error is that he doesn't carefully distinguish among the various types of mental entity that he lumps under the rubric of "Ideas." (Later on the same page as the passage quoted above, Mill criticizes Berkeley for his "ambiguous and misleading use" of the word "Idea.") In one sense, I have an idea of the thing I perceived yesterday even when nobody is perceiving it: I believe or "have the idea" that the thing exists, i.e., according to Mill, that there are certain (unrealized) possibilities of sensation. In another sense, I don't have any idea of the thing when nobody is perceiving it: when nobody perceives the thing, neither I nor anyone else has any of the sensations or "ideas" for which the thing is a possibility.

This interpretation of Mill's passage isn't forced on us by the text. But neither is Hamilton's. Considering that Hamilton's interpretation requires us to see Mill as guilty of a flagrant inconsistency in his position, while the alternative interpretation does not, it seems fair to conclude that the passage doesn't give us a good reason to attribute to Mill the view that permanent possibilities of sensation are beliefs, or the consequent view that physical reality is mind-dependent.

2.4 Legacy issues

For nearly a hundred years, Millian phenomenalism was a significant influence in Anglo-American philosophy, perhaps best known as the unofficial metaphysics of the logical positivists of the early 20th century.[56] As late as 1948, Quine could write: "From among the various conceptual schemes best suited to these various pursuits, one—the phenomenalistic—claims epistemological priority." After that, phenomenalism is notable mainly for its absence from the philosophical scene.[57]

The main reason for the mid-century decline in phenomenalism's fortunes had to do with phenomenalists' use of sensation conditionals.

[56] The best examples of 20th-century phenomenalism are Lewis (1946: 213–17) and Ayer (1946: 169–78). Price (1932) advocates phenomenalism or something very similar. Russell's (1918) position is also recognizably phenomenalist, and there is a strong phenomenalist undercurrent in Carnap (1928).

[57] For Quine's remark, see Quine (1948/1961: 19). The last serious non-exegetical discussions of phenomenalism in the 20th century occurred in the work of early defenders of central state materialism, who saw phenomenalism as a big enough threat to devote substantial effort to discrediting it: see Armstrong (1961: 47–79) and Smart (1963: 16–49).

According to Ayer and other neo-Millian phenomenalists, physical statements are logically equivalent to sensation conditionals or conjunctions thereof. That doesn't mean that it's practical for us to translate any physical statement into an equivalent sensation conditional or conjunction of sensation conditionals. The neo-Millians recognize that the sensational equivalent of even a simple physical statement is probably too complex for us to formulate. However, they also recognize that there isn't necessarily any need to provide actual translations of physical statements into phenomenalistic language, in order to justify the claim that such translations are in principle possible, which is all the phenomenalist equivalence claim commits them to. In this regard, phenomenalists are no worse off than materialists who argue on general theoretical grounds that descriptions of conscious mental states are logically equivalent to descriptions couched in physical terms, without providing any actual translations of mental statements into purely physical vocabulary.[58]

According to the neo-Millians, there's no difference between the existence of some physical thing—say, an apple—and the truth of a corresponding sensation conditional, such as (roughly) that if one were to have experiences as of picking something reddish and roundish off of a tree with certain visible and tangible qualities and then biting into it, one would have apple-flavored taste sensations.

An influential objection to this account appeared in 1948, in the form of Roderick Chisholm's argument from "perceptual relativity." Chisholm argues that sensation conditionals are falsified by the possibility of having perceptual experiences under abnormal conditions: seeing things in peculiar lighting, or in mirrors, or under the influence of drugs. It's possible to have experiences as of picking something reddish and roundish off of a tree with certain visible and tangible qualities and then biting into it, without having apple-flavored taste sensations—due to having a tongue recently injected with novocaine, for example. According to Chisholm, this possibility falsifies the claim that if one were to have experiences as of picking something reddish and roundish off of a tree with certain visible and tangible qualities and then biting into it, one would have apple-flavored taste sensations.[59]

[58] See Ayer (1946: 170–78), and for parallel points about materialist theories, Lewis (1972).

[59] See Chisholm (1948; 1957: 189–97); also Smart (1963: 20–22). Fumerton (1985: 131–50) is a detailed discussion of Chisholm's argument and its aftermath. In effect, Chisholm holds that any sensation conditional that a phenomenalist offers as equivalent to some physical statement has an antecedent that's too weak for the conditional to be true. More recently, Alan Hájek has argued that most counterfactual conditionals are false, due to having consequents that are too

According to Richard Fumerton, Chisholm's argument "was so devastating that singlehandedly it almost cleared the philosophical community of phenomenalists."[60] Judging from the almost complete absence of contributions to the phenomenalist literature in the decades following the publication of Chisholm's argument, that doesn't seem like an overstatement.

Yet it's unclear why we should accept Chisholm's claim that a counterfactual is false if there's a possibility for its consequent to be false while its antecedent is true. In fact, as I'll argue in Chapter 5, we shouldn't accept this claim, given the now-standard analysis of counterfactual conditionals. This alone should go some way towards restoring phenomenalism's fortunes.

However, even if Chisholm's objection fails, there remain unanswered questions. On the standard analysis of counterfactual conditionals, "if it were the case that p, it would be the case that q" means that q is true in the possible worlds most similar to the actual world, among worlds in which it's true that p. If, as I recommend, phenomenalists adopt this analysis, they owe us an account of what makes one world more or less similar to another. Given that they can't appeal to physical similarities—physical features being what they propose to analyze in terms of relevant counterfactuals—it's not obvious what account they can give. Resolving this issue is an important item of unfinished business for phenomenalists.

Phenomenalists' reliance on sensation conditionals is one historical source of dissatisfaction with phenomenalism. Another is the scant detail phenomenalists have offered when it comes to saying exactly what kinds of experiential entities physical things are supposed to be possibilities for. It's clear that they're supposed to be possibilities for experiences to occur in groups like those that idealists identify with physical objects, and it's clear that the experiences in such groups are supposed to relate to each other in the way that experiences typically do relate when they occur in different people perceiving the same physical thing. But what way is that, exactly?

Neither Mill, nor the idealists who preceded him, nor the neo-Millians who came after, give a satisfactory answer to this question. Berkeley refers to experiences "concreted together" to form a physical object; Leibniz compares

strong: see Hájek (2014a). Hájek argues that "if it were the case that p, it would be the case that q" is incompatible with "if it were the case that p, it might not be the case that q," and contends that for most counterfactuals, the corresponding "might" conditional is true. However, even if this is correct, there is, as Hájek acknowledges, a simple work-around: weaken the consequent of the counterfactual conditional so that it reads: "If it were the case that p, then it would *probably* be the case that q."

[60] Fumerton (1985: 141).

the experiences that make up a physical thing to different perspective drawings of the same landscape; Mill speaks of "sensations joined together in groups... [and] so linked together, that the presence of one announces the possible presence at the very same instant of any or all of the rest." Ayer concedes that phenomenalists fail to characterize the relevant groups of experiences other than ostensively, as in "the sort of sense-data that you get when you look at a match-box," but considers this a mainly presentational issue.[61]

The problem with Ayer's nonchalance is that without a clear account of the experiential entities that phenomenalists identify physical things with possibilities for, it's unclear how phenomenalists can account for the intersubjectivity of physical things. This is a serious omission, considering that intersubjectivity is a hallmark of the physical. Giving an account of thinglike groups of experiences that accommodates the intersubjectivity of physical things is another important item of unfinished business for phenomenalists.

[61] Ayer (1946: 170–71).

3
A Signal in the Noise

In this chapter, I develop the argument for phenomenalism roughed out in the early pages of this book. It resembles a well-known argument for materialism, due to David Lewis and David Armstrong. Lewis and Armstrong argue: minds are what explains various regularities in the behavior of certain physical systems (like the bodies of humans and other animals); what explains those regularities are brains having certain physical properties; so, minds are brains with such properties.[1] The parallel argument for phenomenalism is as follows:

1. The physical world is what explains the regularity of experience.
2. What explains the regularity of experience is that there is a propensity for experiences to occur certain ways.
3. Therefore, the physical world is a propensity for experiences to occur certain ways.

Call this the *Regularity Argument* for phenomenalism. It alludes to the "regularity of experience," and to "propensities" for experiences to occur certain ways. Let me explain what I mean by these.

"The regularity of experience" is shorthand for all the prima facie non-random ways that conscious experience unfolds in us, and all the prima facie non-random correlations that hold among different conscious subjects' experiences. To say that experience is regular in this sense is just to say that the experiential facts aren't what you'd expect of a world where experiences occurred completely at random. Experience is often "noisy," but there seems to be a signal in the noise—it's not pure chaos—and that's what I mean by the regularity of experience.

By a "propensity" for experiences to occur some way, I mean a state of affairs we can describe by saying that certain experiences would occur if certain others did, or that the occurrence of certain experiences gives a certain probability to the occurrence of certain others. Basically, propensities

[1] See Lewis (1966) and Armstrong (1968: 82–90).

Phenomenalism: A Metaphysics of Chance and Experience. Michael Pelczar, Oxford University Press.
 DOI: 10.1093/oso/9780192868732.003.0003

for experience are tendencies for experiences to occur certain ways given they occur certain other ways.[2]

The first premise of the Regularity Argument is that the physical world is whatever explains the regularity of experience. This claim is plausible, and, I think, reasonably widely accepted. In §3.1, I argue that it receives support from the way talk about physical things gets into our language.

I then argue that the best explanation for the regularity of experience is that it's a consequence of various conditional facts about experience, understood here as conditional probabilities for the occurrence of one experiential state of affairs given the occurrence of another. I consider three proposed explanations for the regularity of experience: a metaphysical realist explanation (by far the most popular), a Kantian explanation, and a phenomenalist explanation. In §3.2, I argue that the Kantian explanation is superior to the realist explanation, and in §3.3 I argue that the phenomenalist explanation is superior to the Kantian explanation.

Section 3.4 criticizes arguments against bedrock modality, unearthing a further argument for phenomenalism in the process (what I call the Shifting Basis argument).

Section 3.5 concludes with some speculation about the psychological origins of metaphysical realists' resistance to theories that assign chancy facts or counterfacts a central explanatory or metaphysical role.

3.1 Physical reality as explanatory posit

The first premise of the Regularity Argument says that the physical world is an explanatory posit: it's whatever explains why experiences occur in the seemingly non-random ways they do. This premise receives strong, if indirect, support from the fact that experience is our only source of information about the physical world. Experiences occurring in ways that invite explanation are our only evidence of a physical world, and the only reason it occurs to us to describe reality in physical terms. It therefore makes sense for us to use "physical world" as shorthand for "whatever explains why experiences occur these ways (assuming something does explain it)." In this usage, saying that the physical world is what explains such-and-such facts about experience is like saying that gravity is what explains such-and-such facts about motion. People have disagreed about what explains the

[2] I give a detailed account of propensities for experience in Ch. 5.

relevant motions—vortices? a law of attraction? a curvature of spacetime?—but it's generally agreed that whatever best explains the motions wins the title *Gravity*. Likewise, whatever best explains the regularity of experience wins the title *Physical World*—or so the first premise of the Regularity Argument claims.

The claim that the physical world is whatever explains the regularity of experience is compatible with most metaphysical accounts of physical things. If you're a traditional metaphysical realist who thinks that physical things are entities whose natures are exhausted by their geometric and kinematic qualities, you can still agree that the physical world is whatever accounts for the regularity of experience: you'll just say that what accounts for the regularity is something fundamentally geometric and kinematic. If you're a structuralist who thinks physical things are structural features of the world, completely describable in causal and mathematical terms, you can still agree that the physical world is whatever accounts for the regularity of experience: you'll just say that what accounts for the regularity is something structural. If you're an idealist who thinks that the physical things we perceive are mental entities of some sort, you can still agree that the physical world is whatever accounts for the regularity of experience: you'll just say that what accounts for the regularity is something mental.

The first premise of the Regularity Argument is accepted by all who think the right response to external world skeptics is an inference to the best explanation: something explains why experience has the regularity it does (it's not an accident that experience has this regularity); if something explains the regularity, that "something" is the physical world; therefore, there is a physical world. The second premise of this argument is the first premise of the Regularity Argument.[3]

The regularities that anti-skeptics appeal to are typically intrasubjective regularities, since external-world skeptics are apt to regard an appeal to intersubjective regularities as question-begging: according to skeptics, no one has any justification for believing there are other minds. I'm assuming that the skeptics are wrong about this, and that intersubjective regularities in experience are among those that call for explanation—an assumption that no rational skeptic can consistently criticize me for making. In any case, we all

[3] For the abductive argument against external world skepticism, see Locke (1694/1979: IV.xi), Russell (1912: 5), Broad (1925: 140–220), Mackie (1976: 662–9), Jackson (1977: 141–51), Putnam (1982), Vogel (1990), Davidson (2001: 151), Vogel (2005), Chalmers (2010), and Huemer (2015), in addition to the idealists and phenomenalists cited in the next footnote.

do believe that there are other conscious beings who have experiences that relate to our own in prima facie non-random ways. Our thoughts, feelings, and aspirations would be very different otherwise.

I said that the claim that the physical world is whatever explains the regularity of experience is compatible with most metaphysical accounts of physical things. The exceptions are accounts, like Berkeley's, that attribute the regularity to God, understood as a non-physical being. Berkeley's theological concerns fall outside the scope of this study, though I should note that not all theists share them: Spinoza, who holds that God accounts for the regularity of experience, concludes that the physical world is God (or one aspect of God). In any case, even a non-Spinozist theist can agree that for the physical world to exist is for there to be something that accounts for the regularity of experience. If you agree with this much, you can read this chapter as arguing that for there to be a physical world is for there to be suitable propensities for experience.[4]

3.2 "Real things" vs. noumena

The second premise of the Regularity Argument is that what explains the regularity of experience are propensities for experiences to occur certain ways. I argue for this premise in two stages. First, I argue that an explanation in terms of experience-causing powers is superior to one in terms of experience-causing powers grounded in categorical features of "real things"; second, I argue that an explanation in terms of conditional probabilities related to experience is superior to one in terms of the same probabilities grounded in entities with experience-causing powers.

Thus I'm going to consider three proposed explanations of the regularity of experience: a metaphysical realist explanation, according to which what explains the regularities in experience are interactions between conscious subjects and "real things"; a Kantian explanation, according to which what explains the regularities are interactions between conscious subjects and "noumena"; and a phenomenalist explanation, according to which what explains the regularities are the conditional probabilities related to

[4] For Spinoza's metaphysics of *Deus sive Natura*, see Spinoza (1677/1985: I.15). Apart from Berkeley, most mind-first metaphysicians do seem to accept that physical reality is whatever explains the regularity of experience: see, e.g., Leibniz (1705/2013: 325), Kant (1783/1997: §16), Mill (1865/1979: 183–4), Ayer (1936: 53; 1956: 147–8), Foster (2008: 137), and Smithson (2017: 21).

experience that real things and noumena are supposed to ground or underlie. In this section, I argue that the Kantian explanation is superior to the realist explanation; in the next, I argue that the phenomenalist explanation is superior to the Kantian explanation.[5]

In what follows, my focus will be on intersubjective regularities in experience—prima facie non-random relationships among experiences occurring in different conscious minds—since these are the regularities that are most suggestive of a physical world.

Let's start by stating the realist and Kantian explanations of these regularities more explicitly.

The realist explanation is the one most people find most obvious and natural. It goes like this. We have bodies that are embedded in a physical environment. The contents of our physical environment are "real things," by which the realist means, first, that they have *categorical properties,* that is, properties that aren't just tendencies, powers, dispositions, possibilities, propensities, or potentialities—in a word, properties that aren't purely modal—and, second, that they are non-mental (so, no God or monads here). Our bodies, with their non-mental categorical properties, interact with features of our environment, with their non-mental categorical properties, and these interactions often result in our having conscious experiences. It's surprising that such interactions, or any interactions between real things, give rise to conscious experience, but given that they do, it's not surprising that interactions between a given feature of the environment and several similarly-configured observers would cause the latter to have intersubjectively coordinated experiences.

All metaphysical realists attribute intersubjective experiential regularities to interactions between conscious beings and non-mental entities with various categorical properties, but there's disagreement among realists over which properties these are. The most important stances on this question are folk realism and scientific realism.

According to folk realists, the categorical properties responsible for things' experience-causing powers are mainly geometric and kinematic properties, like size, shape, and location, and changes of these over time. For example, an

[5] Two explanations that I'm not going to discuss here are theological explanations, like Berkeley's, and structuralist explanations, according to which what explains the regularities in experience are interactions between conscious subjects and various features of a non-mental structure whose constituents' natures are exhausted by their relations to one another. I discuss structuralism at length in Ch. 7: as that discussion will make clear, the structuralist explanation invites the same criticism as the standard realist explanations criticized below.

object with a certain surface geometry is apt to give us phenomenally green experiences, and an object whose constituent particles are in a certain state of motion is apt to give us phenomenally hot experiences.

According to scientific realists, the categorical properties responsible for things' experience-causing powers are whatever categorical properties feature in ideal completed physics, which might not include traditional "primary" qualities (like size, shape, and motion), and presumably do include some humanly unobservable properties. For example, an object characterized by a certain quantum wave-function, or consisting of subatomic particles with certain spin properties, is apt to give us phenomenally green experiences, or phenomenally hot experiences.

The Kantian explanation is essentially the realist one, minus the requirement that the causes of our experiences be real things (in the realist's sense). There are conscious minds, like ours, and there are entities with powers to cause experiences in conscious minds. The only thing we can know about the latter, which Kant calls "noumena," is that they have various experience-causing powers, some of which they exercise by giving us experiences. We have no basis for attributing the noumena any features beyond these experience-causing powers; in particular, we have no reason to attribute them any categorical properties. Nor do we have any reason to deny or affirm that the noumena are mental.

Realists and Kantians both give causal explanations of the regularity of experience, but they disagree over the nature of the causes. Realists say the causes are non-mental entities with various categorical properties; Kantians suspend judgement on the nature of the causes, beyond ascribing them various experience-causing powers.

Realists account for the regularities in experience by reference to regularities in the categorical natures of non-mental causes of experiences; according to realists, the reason why prima facie coordination between different people's experiences is so common is that it often happens that different people interact with the same real thing.

Kantians account for the regularities in experience by reference to regularities in the experience-causing powers of whatever causes experiences; according to Kantians, prima facie intersubjective coordination arises when different people interact with the same noumenon.

The Kantian explanation is more economical than the realist explanation, since the existence of real things entails but isn't entailed by the existence of noumena: if there are non-mental entities with experience-causing powers grounded in their categorical properties, then there are entities with experience-causing powers, but the reverse doesn't hold. The Kantian's ontological commitments are therefore a proper subset of the realist's. From a Kantian standpoint, realism is just the known truth embellished by a conjecture about the ultimate origins of our experience.[6]

The parsimony of the Kantian explanation is an important point in its favor. Other things being equal, a more parsimonious explanation is preferable to a less parsimonious one. Are other things equal here?

They are, provided that (1) the realist explanation explains no more than the Kantian explanation, and (2) the realist explanation is no better at making predictions than the Kantian explanation. If the Kantian, whose terms are a proper subset of the realist's, can explain everything the realist can explain, and predict everything the realist can predict, the Kantian has a better theory.[7]

The realist account does not explain more than the Kantian account. Suppose a realist says that the reason why there's a certain experiential regularity R is that there are objects with certain geometric and kinematic features ϕ. Then a Kantian can say that what explains R is that something has (or some things have) the power to cause experience as of objects with features ϕ.

A realist might argue that his account explains regularities in the causes of our experiences, and not just regularities in our experiences. For example, the realist account explains why a flying baseball breaks a window: it's because interactions between (1) things with the primary qualities of flying baseballs and (2) things with the primary qualities of windows tend to result in (3) things with the primary qualities of broken windows. In the realist

[6] Michael Devitt rejects the Kantian explanation on the grounds that a world of Kantian noumena is "a world we cannot know about or talk about. It cannot play a role in explaining any phenomenon... [I]t does not require the existence of any familiar objects, but simply a Kantian noumenal world of 'things-in-themselves'" Devitt (1991: 17). But a Kantian says that we *can* know about a world of noumena (through an inference to the best explanation), that the noumena *do* play a role in explaining many phenomena (namely, the regularities in our perceptual experience), and that the existence of noumena *does* require the existence of familiar objects, since for familiar physical objects to exist just is for there to be noumena with suitable experience-causing powers.

[7] I take it that scope (amount explained), parsimony (economy of terms), and predictive success are the most important virtues for an explanation to have. So, if theory A is superior to theory B in at least one of these "big three" respects, and inferior to B in none of them, A is preferable to B, regardless of the theories' subordinate explanatory virtues (like "elegance" and "unity")—not that realists necessarily have an advantage over Kantians when it comes to subordinate virtues. Of course, the paramount virtue for an explanation to have is truth, but since we're trying to find out which explanation is true, we can't appeal to truth as a deciding factor.

account, things' primary qualities explain not only the regularities in their experiential effects on us, but regularities in their non-experiential effects on one another. (For a scientific realist account, substitute "ideal scientific" for "primary.")

Notice, first of all, that a Kantian can give an explanation that parallels the one the realist is offering here. A Kantian can say that a flying baseball breaks a window because interactions between (1) entities with the power to cause experiences of things with the primary qualities of flying baseballs and (2) entities with the power to cause experiences of things with the primary qualities of windows tend to result in (3) entities with the power to cause experiences of things with the primary qualities of broken windows. Noumenal experience-causing powers explain not only the regularities in the noumena's experiential effects on us, but regularities in how the noumena affect each other's experience-causing powers.[8]

Second, and more basically, notice that the "extra" regularities realists claim to explain are really just further regularities in experience. The existence of observed regularities in the causes of our experiences is nothing but the existence of regularities in the experiences that those supposed causes supposedly cause. At least, the latter regularities are our only reason for believing in the former.

The Kantian explanation of the regularity of experience is more economical than the realist explanation, and, I've just argued, no less comprehensive than the realist explanation. How do the explanations compare in terms of predictive success?

There is a general reason to doubt that realism has any predictive advantage over Kantianism. Ultimately, the inputs to and outputs of the models that generate predictions are experiences: ultimately, to make a prediction is to tell us what experiences to expect, given other experiences. Well, systems that give the same outputs for the same inputs have equal predictive success. Since Kantians can assign experiential outputs to experiential

[8] You might wonder whether Kantians incur a disadvantage by ascribing experience-causing powers to unobserved primary or ideal scientific features of things where realists do not. Below, I argue that Kantians don't incur a disadvantage, since we wouldn't believe atoms existed if we didn't think it was possible to perceive an atom—perceivability being what distinguishes a physical thing from a mathematical theorem, or a fictional character, or a dream. Even if Kantians were at a disadvantage here, it would only be with respect to a subordinate virtue, like elegance. (But perhaps with a compensating advantage in unity?) The Kantian theory would still be more parsimonious than the realist theory in the way that matters most: it would still require fewer fundamental kinds of thing, in fact a subset of the kinds realism requires: the realist kinds, minus non-mental categorical properties and non-mental entities to instantiate them.

inputs however realists can, it's hard to see how Kantians can ever be at a disadvantage compared to realists when it comes to making predictions.

For example, suppose a folk realist explains the chromatic regularities in our experiences of monarch butterflies by reference to the light-scattering effects of the surface geometry of the butterflies' wings, and on this basis predicts that people who see viceroy butterflies, whose wings have a surface geometry similar to that of monarchs, have experiences similar in phenomenal color to those they have when they see monarchs. Then a Kantian can explain the chromatic regularities by reference to entities empowered to cause experiences of surfaces with a certain geometry imperceptible to the naked eye, and predict on that basis that the entities empowered to cause experiences of viceroys, which have the power to cause experiences of surfaces with the same geometry, give the people they affect experiences chromatically similar to those we have when seeing monarchs. Whatever laws relate objects with various surface geometries to experiences with corresponding phenomenal colors can also be made to relate entities empowered to cause experiences of objects with those surface geometries to experiences with the corresponding phenomenal colors.

Here it's fair to ask what is to be gained by attributing experience-causing powers to microscopic phenomena like surface geometries. If folk realists can put the surface geometries to use without supposing that they have any particular experience-causing powers, isn't the realist account more conservative than the Kantian?

What is to be gained by attributing experience-causing powers to microscopic phenomena, says the Kantian, is the right to say that such phenomena exist. An ability to reveal itself to conscious beings through perceptual experience—an ability to be perceived—is what distinguishes a real physical thing from a fiction, abstraction, or phantasm. According to Kantians, we assume the experience-causing powers when we assume the microscopic phenomena. That's why if microscopic investigation revealed no variation in the surface geometries of wings of different-colored butterflies, we'd have to abandon the surface-geometry explanation of the chromatic regularities in our experiences of butterflies.

The story is much the same when it comes to scientific realism. If a scientific realist explains the chromatic regularity in our experiences of monarch butterflies by reference to particles with humanly unobservable quantum properties, a Kantian can explain it by reference to entities empowered to cause non-human experiences of particles with those properties. If you ask what justifies attributing experience-causing powers to quantum states, the answer is: whatever justifies believing that quantum states are more than a useful fiction.

Maybe it's not so clear that we can speak of perceiving a quantum state the way we can speak of perceiving a macroscopic object or microscopic surface feature. But if there's any doubt about the propriety of such talk, that's a reason to question the reality of quantum states, and to suspect that talk about quantum states plays a non-referential role in scientific prediction-making—in which case Kantians are free to engage in such talk too.[9]

Before going any farther, let me sort out an ambiguity about imperceptibility or unobservability. In one sense, to say that something is imperceptible or unobservable is to say that no actual sentient being is able to perceive it; something unobservable in this sense is what we might call *contingently unobservable*. In another sense, to say that something is unobservable is to say that it's metaphysically impossible for there to be perceptions of it; things unobservable in this sense are what I'll call *metaphysically unobservable*.

Kantians have no problem with contingently unobservable things: they're just things with powers to cause experiences that no actual being is able to have, due to contingencies of evolution, resource constraints, natural laws, or whatever. For example, suppose that atoms are contingently unobservable. Then, like everything we know through experience, atoms have features and behave in ways that we can't in fact perceive, but that we might have been able to perceive if we'd had sufficiently sensitive sense organs, and that some alien or machine sentience might actually perceive now. In this regard, atoms are no different from sticks and stones: the experiences through which we know them are a small, unrepresentative sample of those they have the power to cause, all of which we'd have to take into account to get a full picture of any given physical thing (stone, atom, quantum state, or whatever).

Metaphysically unobservable physical things are a different story: according to Kantians, such things are impossible. Whether or not Kantians are right about this, they can't be proved wrong, as I now argue.

[9] The view that scientific models constructed out of quantum states yield accurate predictions despite there not being such things as quantum states is a kind of scientific antirealism. According to scientific antirealists, quantum-state talk is just a useful kind of non-descriptive discourse, and quantum states are useful fictions, like Bohr's atoms or Newtonian spacetime. (Actually, most self-styled scientific antirealists are scientific agnostics, suspending judgement on the existence of things like quantum states rather than positively denying that they exist; see, e.g., Duhem 1914 and van Fraassen 1980.) I take no stand on this here; however, for the reasons I give below, anti-realism does seem like a plausible stance to take towards any physical entity purported to be not just humanly imperceptible, but imperceptible in principle.

Take someone who has up to this point been an antirealist about a certain kind of thing that he's never perceived. What would it take to rationally compel him to change his mind and become a realist about that kind of thing?

One thing that would be enough would be actually perceiving such a thing. If you're a germ skeptic, and then someone shows you germs under a microscope, you can't reasonably continue to be a germ skeptic.

Another would be acquiring credible evidence that other people perceive the relevant kind of thing. People blind from birth acquire this kind of reason to believe that there are stars, reflections, rainbows, etc.

A final thing that would be enough to rationally compel a conversion to realism would be learning that the relevant kind of thing, though never in fact perceived by anyone, is the kind of thing one could perceive. The fossil record gives us this kind of reason to believe there were single-celled organisms on Earth 3.5 billion years ago.

First-hand experience of something, learning that others have experienced it, and learning that the relevant kind of thing, though never in fact perceived, is perceivable: these are what can compel a rational antirealist to convert to realism about something whose existence he previously doubted. There might be other reasons to convert, but they're the sort of reasons whose probative value reasonable scientific realists and antirealists disagree about: the explanatory or predictive value of assuming that the disputed entity exists, the scientific track record, etc. As long as the antirealist never perceives *x*, never learns that others perceive *x*, and never learns that *x* is the kind of thing that while never in fact perceived is perceivable, he has no compelling reason to stop suspending judgement on *x*'s existence.

What about a metaphysically unobservable physical entity? Since there's no possibility of perceiving it, no antirealist about it can perceive it. So, an antirealist about it can't be rationally compelled to adopt a realist stance on it by experiencing it first-hand. And since there's no possibility of perceiving it, an antirealist about it can't be rationally compelled to adopt a realist stance on it by learning that others perceive it or could perceive it. Since these are the only ways to rationally compel a scientific antirealist to convert to realism, there's no compelling reason to be a scientific realist about any metaphysically unobservable physical entity: the point at which it becomes reasonable to doubt that there's a possibility for experience of some alleged physical entity is the point at which it becomes reasonable to adopt an attitude of scientific antirealism (or agnosticism) towards that alleged entity.

This means that scientific realists can't legitimately claim to have access to scientific resources unavailable to Kantians. If realists can build their models with atoms understood as actual physical entities and not just useful fictions, Kantians can build their models with entities empowered to cause suitably regular experiences of atoms. On the other hand, if realist talk about atoms serves its purpose without referring to anything, it can serve the same purpose for the Kantian.

3.3 Noumena vs. possibilities of sensation

I've argued that the Kantian explanation of the regularity of experience is superior to realist explanations (whether in terms of traditional primary qualities, as in folk realism, or ideal scientific properties, as in scientific realism). In this section, I argue that a phenomenalist explanation of the regularity of experience is superior to the Kantian explanation.

Let's review what it is we're trying to explain. We're trying to explain why experiences, such as those we have in ordinary perception, have certain seemingly non-random features, and occur in seemingly coordinated ways, and, in particular, why there are seemingly non-random similarities and correlations across different conscious subjects' mental lives. These regularities take the form of seemingly non-random relative frequencies among perceptual experiences. When we're playing frisbee in the park, the frequency of disc-like and park-like visual impressions in me relative to disc-like and park-like visual impressions in the people I'm playing frisbee with is greater than chance (or so it seems). Likewise, it seems to be not just by chance that as I have experiences as of a successively more distant disc, you have experiences as of a successively nearer disc. The question is how best to explain these and other, similar experiential regularities.

Kantians and metaphysical realists agree that there are objective probabilities for experiences to occur in certain ways conditional on their occurring in certain other ways, since they agree that the seemingly non-random patterns in experience are genuinely non-random. According to metaphysical realists, these probabilities are grounded in categorical features of the things that cause experiences; according to Kantians, the probabilities are grounded in experience-causing powers of the things that cause experiences, though we have no basis for attributing any categorical properties to those things.

The phenomenalist explanation is essentially a streamlined version of the Kantian. In the Kantian view, what explains the regularity of experience are entities with suitable experience-causing powers. According to phenomenalists, what explains the regularities are propensities for experiences to exhibit the regularities. The Kantians' noumena with their experience-causing powers wouldn't explain the regularities in our experience, if they didn't ground objective probabilities for experience to exhibit such regularities. But we can appeal to the probabilities without positing any noumena to ground them. So, per Ockham, we shouldn't posit any noumena. We should make do with the probabilities alone.

This is the phenomenalist explanation I favor: the reason why experience seems to occur non-randomly is that it *does* occur non-randomly; experience owes its prima facie non-randomness to the fact that there are objective probabilities for experiences to occur in certain ways, given that they occur in certain other ways.

To show that the Kantian explanation was superior to realist explanations, I argued that the Kantian explanation was more parsimonious than realist explanations, explained at least as much as realist explanations, and had predictive success equal to that of realist explanations. As we've just seen, the phenomenalist explanation is more parsimonious than the Kantian explanation. To establish its superiority to the Kantian explanation, it remains to show that the phenomenalist explanation is at least good as the Kantian in terms of how much the theories explain, and how well they make predictions.

Phenomenalism explains no less than Kantianism. Kantians posit noumena with experience-causing powers to explain the regularity of experience. As argued in the previous section, a comprehensive explanation of the regularity of experience entails an explanation of regularities among the things we experience: ultimately, explaining the latter regularities is just a matter of explaining regularities in the experiences by which we observe them. Since both phenomenalists and Kantians offer comprehensive explanations of the regularity of experience, the explanations have the same scope.

Phenomenalism is no less predictive than Kantianism.

Suppose a Kantian predicts that someone who sees a viceroy butterfly will have experiences similar to those of someone who sees a monarch butterfly, by saying that in both cases, the experiences arise from interactions between conscious subjects and noumena empowered to cause experiences of surfaces with a certain geometry imperceptible to the naked eye. Then a phenomenalist can make the same prediction, on the grounds that the probability of having phenomenally orange and black experiences, given that one has experiences as of viewing a surface with a certain geometry under a microscope and then viewing the same surface with the naked eye, is high. If there's a lawlike relationship between entities empowered to reveal themselves via experiences of objects with certain surface geometries and entities empowered to cause experiences with certain phenomenal colors, then there's a lawlike relationship between possible experiences as of objects with those surface geometries and possible experiences with those phenomenal colors.

What about scientific predictions based on the behavior of humanly unobservable phenomena? The phenomenalist view of these predictions is similar to the Kantian. If we have good reason to accept the humanly unobservable posits of science as genuine entities (rather than useful fictions), then the posits must come with associated possibilities of sensation. If atoms, quantum states, etc. don't support the type of experience-related counterfactual conditionals or conditional probabilities that we expect physical things to support, that's a reason to doubt that there really are such things as atoms and quantum states.

3.4 The superfluity of Substance

I've argued that the phenomenalist explanation of the regularity of experience is better than realist and Kantian explanations. However, there's a view in the metaphysics of modality that, if correct, might seem to weigh heavily against the phenomenalist explanation, or even against phenomenalism itself: the view that it's impossible for any modal fact (about possibilities, probabilities, potentialities, or whatever) to exist in the absence of underlying non-modal facts—that it can't be modality all the way down.

David Armstrong defended a version of this view throughout his long career. According to Armstrong, a modal state of affairs never exists except in virtue of some non-modal (categorical) state of affairs. Call this the *Armstrong doctrine.*[10]

By itself, the Armstrong doctrine doesn't pose a direct threat to phenomenalism, which doesn't assert that possibilities of sensation lack categorical grounds, but merely refrains from asserting that they do. However, you could try to use the Armstrong doctrine to undermine the Regularity Argument for phenomenalism. In the spirit of Armstrong and Smart, you might argue that when some phenomenon has a categorical explanation, the categorical

[10] See Armstrong (1961: 85–8; 1993: 187), and his contributions to Armstrong et al. (1996). Armstrong's view is also known as categoricalism.

explanation is always superior to a non-categorical explanation of the same phenomenon; in particular, you might argue, if some categorical entity explains the regularity of experience, it's that categorical entity that best explains the regularities, rather than the phenomenal possibilities that the categorical entity grounds.[11]

The claim that a categorical explanation is always better than a non-categorical one is contentious, and (I argue below) false. Setting that aside for now, a more basic problem with the foregoing argument is that we have no good reason to accept the Armstrong doctrine. If anything, we have good reasons to reject it.

The Armstrong doctrine isn't widely accepted these days. This is mainly due to an increased awareness among philosophers that it's an open scientific question whether a certain class of modalities—the probabilities associated with quantum mechanics—have any categorical basis. According to some interpretations of quantum mechanics they do, according to others they don't; we'll survey these interpretations later, in Chapter 6. The important point here is that the scientists wouldn't consider this an open question if they thought it was logically impossible for modalities to exist without a categorical basis. The consensus view, among both scientists and philosophers, is that it's up to physics, not metaphysics, to decide whether quantum probabilities have a categorical basis.

Maybe the scientists are wrong not to think that modalities require a categorical basis, but their skepticism has this in its favor: it's easy to imagine categorically identical possible worlds that differ in their probabilities.[12]

For example, there's about a 50% chance that a radon atom decays within four days of the time it comes into existence. Presumably, about 50% of the radon atoms that come into existence in our world decay within four days. But we can imagine a world categorically indistinguishable from ours, in which radon atoms have a 90% chance of decaying within four days. We only need to imagine that due to a statistical fluke, about 50% of the radon atoms in this other possible world decay within four days, despite there being a 90% chance of any given radon atom decaying within four days. In this other world, there exist probabilities (for radioactive decay) that don't exist in our world, despite the two worlds being identical in all categorical respects.

[11] See Armstrong (1961: 56–8) and Smart (1963: 22–3). Armstrong and Smart hold that whenever there's a non-categorical explanation for something, there is a deeper categorical explanation.

[12] The point isn't new: see McKitrick (2003) and Mumford (2006).

Assuming that what we're imagining is logically possible, it follows that the probabilities for radon decay that exist in our world don't exist in virtue of our world's having some categorical feature.

You might question whether what we're imagining here really is possible. That's a legitimate question to raise about any conceivability argument, of which the above is an example. But the burden is on those who raise it to provide a specific reason to think that our modal intuitions are unreliable in this case, and it's not obvious what that reason might be.

So much for reasons to suspect that the Armstrong doctrine is false. Is there any reason to think the doctrine is true?

Armstrong argues that we have to accept that dispositions and other modalities require a categorical basis, in order to avoid committing ourselves to non-existent entities. According to Armstrong, if dispositions lack any categorical basis, then when an object has an unmanifested disposition, "*the object still has within itself, essentially, a reference to the manifestation that did not occur*. It points to a thing that does not exist." This is objectionable, says Armstrong, because it entails "a Meinongian metaphysics, in which actual things are in some way related to non-existent things."[13]

The idea, apparently, is that if a thing *x* has a disposition to do *V*, *x* must point to *V*; so, if *x* has an unmanifested disposition to do *V*, and this disposition has no categorical basis, then *x* must point to *V* by pointing to something that is neither *V* itself (since the disposition is unmanifested) nor any categorical entity. And from this it supposedly follows that if *x* has an unrealized and categorically unbased disposition, there must be some non-existent entity that *x* points to.

It's unclear why something with a disposition needs to "point to" whatever it has a disposition to do; it's not even clear what this means. Anyway, we can hold that some dispositions are categorically ungrounded without talking about pointing. A thing has a categorically ungrounded disposition when a certain conditional involving it is true (maybe a counterfactual conditional, maybe a conditional probability—the details aren't important here), and the conditional either has no basis in anything or has a basis only in other conditionals. When something with a categorically ungrounded disposition fails to manifest the disposition, that just means that the antecedent of the conditional is unsatisfied. There's no commitment to Meinongian metaphysics in any of this, any more than in the claim that "if it were the

[13] Armstrong (1997: 79).

case that p, it would the case that q" or "the probability that q, given that p, is high" can be true, even when it's false that p.[14]

A different argument for the Armstrong doctrine comes from Michael Smith and Daniel Stoljar. According to them, the doctrine follows from a plausible analysis of statements of the form

> *x* is disposed to verb in circumstance C at possible world *w*.

as equivalent to:

> *x* verbs in all the possible worlds most similar to *w*, among those worlds in which circumstance C holds.

If this analysis is correct, it's impossible for something to have a disposition except in virtue of some further feature of the world that makes it more similar to some C-worlds than others.[15]

Proponents of the view that there are categorically ungrounded dispositions can accept this analysis and its implications. They can hold that the worlds most similar to *w* among those in which C holds include worlds that are exactly the same as *w* in all categorical respects, differing from *w* only in what dispositions exist in them. The analysis of dispositions that Smith and Stoljar deploy entails that all dispositions are grounded in something, but it doesn't entail that all (or any) dispositions are grounded in something besides further dispositions.[16]

There's a related analysis of "*x* is disposed to verb in circumstance C at possible world *w*" that would, however, threaten the view that there can be ungrounded modalities. According to this related analysis, the statement that *x* is disposed to verb in C at *w* is equivalent to: "*x* verbs in all the possible worlds most similar to *w* in *non-dispositional* respects, among those worlds in which circumstance C holds." Unlike the earlier analysis, this one

[14] The exact relationship between dispositions and conditionals is controversial, but whatever it may be, the modalities expressed by counterfactual conditionals and statements of conditional probability certainly escape Armstrong's objection, for the reasons given.

[15] See Stoljar and Smith (1998: 90–99).

[16] Curiously, this is a point that Stoljar and Smith concede (1998: 94). They say that a view in which dispositions are grounded only in further dispositions isn't a "bare dispositions" view, since it grounds dispositions in something (albeit only other dispositions). Be that as it may, the view is certainly at odds with the claim that "if the world is a certain way dispositionally it must also be a certain way in and of itself: that is, a certain way non-dispositionally" (Stoljar and Smith 1998: 87). So the proposed analysis of disposition statements doesn't compel us to posit categorical grounds for dispositions or other modalities.

does conflict with the view that there could be categorically ungrounded dispositions.[17]

It's question-begging to argue for the Armstrong doctrine by appealing to this new analysis of dispositions. A proponent of categorically ungrounded dispositions should simply reject the analysis, as incompatible with the intuition that two worlds could duplicate one another in their non-dispositional features yet differ dispositionally, as in the case of the actual world and a categorically indistinguishable world in which radon has a shorter half-life than it has in the actual world.[18]

The prima facie threat that the Armstrong doctrine posed to phenomenalism was that it guaranteed the existence of a categorical explanation for the regularity of experience, where the existence of such an explanation allegedly makes it more reasonable to identify the physical things we perceive with something categorical, rather than with the possibilities of sensation that this categorical something grounds. However, the latter allegation is doubtful, for the following reason.

Imagine that up to a certain point in time, the tendency for our perceptual experiences to exhibit the regularities they do is grounded in a system of Leibnizian monads. Then, some time later, and without any phenomenological consequences, the tendency loses its monadic basis and acquires a new basis in God's mind, à la Berkeley. Shortly after that, the tendency loses its divine basis and acquires a new basis in a supercomputer (as in a Matrix scenario), etc.[19]

If the physical things we perceive are the assumed categorical bases of our experiences' propensities to exhibit the regularities they do, rather than the propensities themselves, the shift from a monadic to a divine to a computational basis has the following implication: when you say, pointing to the Moon: "That's the same one we saw yesterday," you speak falsely. This is implausible: as implausible as the idea that you speak falsely if, pointing to a waterfall that you're visiting for the second time, you say that it's the same one you saw before.

Is this taking too narrow a view of what counts as the ground or basis of something? Could you say that in a monads-to-God scenario, what

[17] See Stoljar and Smith (1998: 96–9), arguing that such an analysis is justified by the alleged fact that any difference or similarity between two possible worlds must be a difference or similarity in the worlds' non-dispositional features.

[18] Let x = a sample of radon atoms, verb = reduce by half, C = four days have elapsed since the sample was created, and w = the actual world.

[19] Robinson (1994: 222–6) considers a similar thought-experiment.

grounds the relevant possibilities of sensation—the relevant propensities for our experiences of the Moon to exhibit the regularities they do—is a disjunctive entity that consists of monads up to a certain time, and God's mind thereafter?

You could. But if you're willing to say that the same moon can exist at one time as something that grounds certain possibilities of sensation, and at another time as something else that grounds the same possibilities of sensation, then it seems that what matters to you as far as the question of the Moon's existence is concerned is the possibilities of sensation, not whatever might happen to ground them. And the possibilities do seem to be what matters to us. If we're planning a mission to the Moon, we're counting on its being highly probable that the astronauts will have experiences as of approaching the Moon if they have experiences as of travelling a certain direction for a certain period of time at a certain speed. If we thought the probability would sharply decrease halfway through the mission, we wouldn't approve the launch. But if we're confident that the probability will remain high (along with other, related probabilities), we won't care whether this is due to monads, God, or any other categorical feature of the world. Nor will we care whether what keeps the probability high is a single underlying entity or a succession of distinct underlying entities.

If the Armstrong doctrine is true, it follows that the regularities in our experience have an explanation in categorical terms. But the Armstrong doctrine doesn't imply, and we have no other reason to believe, that the regularities in our experience have any *stable* basis in some categorical feature of the world. We have evidence that the regularities have a stable basis in *something*—frisbees, moons, and so forth—but our empirical evidence doesn't support the hypothesis that the basis is something categorical, and even if we had a priori reasons to think that everything must have some categorical basis or other—that it can't be possibilities all the way down—we'd still have no reason to think the regularities in our experience had diachronically stable categorical bases. Yet we know that we often perceive the same physical thing at different times. So even if the Armstrong doctrine were true, we'd still have a reason to identify the physical things we perceive with corresponding possibilities for experience, rather than with whatever categorical phenomena might ground the possibilities.

This gives us an additional argument in favor of preferring phenomenalism to metaphysical realism. Call it the *Shifting Basis Argument:*

1. If we know that some physical things persist, then physical things are possibilities of sensation rather than the categorical entities, if any, that ground the possibilities.
2. We know that some physical things persist.
3. Therefore, physical things are possibilities of sensation rather than the categorical entities, if any, that ground the possibilities.

A similar argument gives us an additional reason to prefer phenomenalism to Kantian metaphysics. Unlike metaphysical realists, Kantians don't assume that the causes of our experiences have categorical properties, but Kantians agree with metaphysical realists in denying that it's modality all the way down. According to Kantians, experience-causing powers must be powers *of* something, and this "something" can't just be more powers. There must be non-modal entities that possess the powers. Call this the *Kant doctrine.*

Unlike the Armstrong doctrine, there are no arguments for the Kant doctrine, as far as I know.[20] But even if the doctrine is true, it doesn't threaten phenomenalism. By Kant's own lights, the only thing we can know about noumena is that they're what causes experiences. There's no way we can know whether some experiences occurring today are caused by the same noumenon as some other experiences that occurred yesterday. So, if we were to equate the existence of a physical thing, like the Moon, with some noumenon's possession of suitable experience-causing powers, we'd have to conclude that we can never know whether the Moon exists from one moment to the next. Since we do know this, we should reject the proposed equation. This gives us the following argument to prefer phenomenalism to Kantian metaphysics:[21]

1. If we know that some physical things persist, then physical things are possibilities of sensation rather than the noumena, if any, whose experience-causing powers ground the possibilities.

[20] The closest Kant comes to giving an argument for it is when he says that for there to be appearances, there must be something that appears: Kant (1781/1998: 115). Taken as an argument for the Kant doctrine, this is feeble, as becomes obvious when you replace "appearances" with "experiences."

[21] This is similar to Mill's argument for the superfluity of "Substance," understood as something that somehow grounds or supports possibilities of sensation: see Mill (1865/1979: 198), quoted above on pp. 21–22.

2. We know that some physical things persist.
3. Therefore, physical things are possibilities of sensation rather than the noumena, if any, whose experience-causing powers ground the possibilities.

3.5 The ultimate sin in metaphysics?

I've argued that the regularity of experience is attributable to objective conditional probabilities related to experience. Many people are uncomfortable with this conclusion. More generally, people tend to be uncomfortable with explanations based on probabilities, and uncomfortable too with assigning probabilities a foundational metaphysical role. What is the source of their discomfort?[22]

I'm not sure, but I suspect it's due to a known tendency for people to distrust appeals to probability (or, more generally, modality) even when their distrust has no rational basis.

For example, it's a fact well known to trial lawyers that people are reluctant to return guilty verdicts on the strength of explicitly statistical evidence alone, despite the fact that the so-called "direct" evidence that people prefer is, at bottom, itself statistical. If you're spat on by someone you vividly remember as a large Caucasian man, and a DNA test indicates the origin of the saliva as a petite Asian woman, you'll have a hard time accepting the DNA evidence over the testimony of your own visual memory, even if you're aware that eyewitness identification has a much lower accuracy rate than DNA matching, and you know that the DNA test was properly conducted.[23]

We all know the saying "Seeing is believing," with its connotation that sight gives us a kind of justification that transcends anything we can get through an inference from information about mere likelihoods. We tend to forget that our confidence in the reliability of sight comes from our knowledge that the incidence of true sight-based beliefs relative to all sight-based beliefs is high, and our belief that this isn't accidental, but reflects an objective likelihood

[22] Some reactions go beyond discomfort. According to David Armstrong, "letting dispositions 'hang on air'... is the ultimate sin in metaphysics" (2004: 3). It's clear he would say the same about probabilities, powers, propensities, or any other form of modality.

[23] See Schoeman (1987) and the references therein, particularly Nisbett and Ross (1980).

for sight-based beliefs to be true. We tend to forget that eyewitness evidence is also statistical evidence.[24]

Or consider the way people reacted when national governments went off the gold standard in the 1930s. Many people felt that their money lost value as a result of this move. Even though very few people had ever actually exchanged currency for government gold, many people thought that the value of government-issued paper money depended on the government's promise to redeem currency for gold on demand.

What those people failed to realize was that the value of gold is itself based on a promise: not an explicit promise between the state and its citizens, but an implicit promise among citizens to accept gold as a medium of exchange. (There is ornamental and industrial demand for gold, but nearly all of gold's value lies in demand for it as an exchange medium.) Ultimately, the value of gold has the same basis as the value of currency unbacked by any commodity: a disposition for people to accept it in exchange for other goods and services.[25]

I conjecture that people's discomfort with the phenomenalist explanation of the regularities in experience, and their preference for realist explanations, have the same psychological source as their preference for eyewitness evidence over explicitly statistical evidence, or for gold-backed currency over fiat currency. Nothing in this chapter's argument relies on this conjecture, but if you have strong anti-phenomenalist intuitions, I hope that seeing the analogy between them and other intuitions known to be unreliable will make you more receptive to the phenomenalist worldview.

[24] Likewise, as Schoeman points out (1987: 185), while fingerprints are typically regarded as a form of direct evidence, the reason we have so much confidence in fingerprint evidence is the low probability of more than one person having the same print-type.

[25] Money's basis in human dispositions or "force of habit" has been recognized at least since Jevons (1875); see also Kiyotaki and Wright (1991). Warren Buffett estimates that for the cost of buying the world's entire gold supply (which would fit into a 68 ft cube) one could purchase all U.S. cropland plus 16 ExxonMobils, with $1 trillion left over. Whatever intrinsic economic value a 68 ft cube of gold may have, it's certainly far less than that of 400 million acres of farmland.

4
Possibilities for What?

I've said that phenomenalists see the physical world as the possibility for things to be as traditional idealists say they actually are. This would be more illuminating, if traditional idealists said more than they do about how they think things are, and phenomenalists more about what they mean by saying that there's a possibility for things to be that way. The purpose of this chapter is to say exactly what I take the physical world to be a possibility for. (In the next chapter I'll tell you what I mean by saying that there's a possibility for it.)

At the end of Chapter 2, we saw that phenomenalists face a challenge accounting for the intersubjectivity of physical objects and events. I begin the present chapter by reviewing this challenge to phenomenalism. My response to the challenge hinges on developing a notion of ideal spacetime, modelled on physical spacetime as understood in current mainstream physics. In §4.2, I lay out the basic resources required to define physical spacetime, and in §4.3 I define ideal spacetime in terms of parallel experiential resources. Finally, in §4.4, I define thinglike groups of experiences as groups of experiences co-located in ideal spacetime, and ideal worlds as totalities of thinglike groups existing in the same ideal spacetime. These definitions sharpen and supersede the definitions of ideal worlds and thinglike groups given earlier.

4.1 The challenge of intersubjectivity

Earlier I described ideal observers who perceive the physical world without affecting it in any way. I asked you to imagine that there are many ideal observers: so many that they collectively perceive every physical feature of our universe. The ideal observers' experiences stand to each other in various relations of counterfactual or probabilistic interdependence, like those that relate actual people's experiences when they perceive the same physical object or event. This hypothetical network of experiences is an example of what I've called an ideal world.

As noted earlier, there is, for every possible physical world, a corresponding ideal world, consisting of the network of experiences that would exist if

Phenomenalism: A Metaphysics of Chance and Experience. Michael Pelczar, Oxford University Press.
 DOI: 10.1093/oso/9780192868732.003.0004

that physical world were populated by ideal observers as described above. Let's call the ideal world that corresponds to a given possible physical world the *ideal counterpart* of that physical world.[1]

For most of the world's physical features, it's relatively easy to find a counterpart in the corresponding ideal world. For physical shapes, sizes, masses, charges, velocities, and other properties, there are the sort of experiences that we normally describe in terms of shape, size, mass, etc., as when you describe your visual experience of a penny as (phenomenally) round, or your experience of a bowling ball as (phenomenally) heavy. There are nuances here related to the diversity of sensory modalities by which different conscious beings experience the world, and we'll discuss these in due course. But, to a decent first approximation, the phenomenalist view is that the ideal counterpart of a physical thing is a group of experiences like that which typically exists when multiple people perceive the same thing.

There's one important feature of physical things, though, that doesn't have any immediately obvious ideal counterpart: their intersubjective accessibility. This is one of the main things that distinguishes physical entities from conscious experiences. I can, in a sense, perceive my own visual experiences: I can be introspectively aware of them. But nobody else can perceive them, by introspection or any other means. Visual impressions of flames, unlike flames, are not intersubjectively accessible.[2]

Intersubjective accessibility is arguably the most important criterion of physical thinghood. It's therefore imperative for the phenomenalist's possibilities of sensation to be intersubjectively accessible. Unfortunately, as we saw in Chapter 2, this is a place where historic phenomenalists (as well as traditional idealists) have stumbled.

Mill seems to think that it's enough for intersubjective accessibility if possibilities of sensation are possibilities for groups of suitably interdependent experiences occurring in different minds. But as the following example shows, this is at most necessary, and not sufficient, for intersubjectivity.

[1] For ease of exposition, I assume that every physical world has just one ideal counterpart. In effect, this is a choice to require a physical world's ideal counterpart to include the experiences of every possible type of ideal observer: observers who perceive things via experiences like ours (visual, tactile, etc.), observers who perceive things via experiences similar to but distinct from ours (e.g., chromatic inverts), observers who perceive things via experiences in sensory modalities utterly alien to us, etc. I make this choice only for convenience's sake: we could equally speak of the ideal counterparts (plural) of a given physical world, and identify our physical world with the possibility for at least one of its ideal counterparts.

[2] Unless materialism is true—a possibility I set aside for now. The important point here is that physical things are intersubjectively accessible, not that conscious experiences aren't.

Suppose there are two planets, Earth and Twin Earth, that are perfect physical duplicates. Suppose furthermore that due to the initial conditions of the universe, Earth and Twin Earth evolve in perfect parallel, so that whenever a physical change takes place on either planet, an identical change takes place on the other. If some people on Earth are standing around a flagpole watching a flag waving in the wind, some physically identical people on Twin Earth are watching a physically identical flag undergoing identical motions.

There is interdependence among the Earthlings' experiences—the Earthlings' flag-experiences co-evolve like the Bedouins' experiences described earlier—and there's interdependence among the Twin Earthlings' experiences. But there's also interdependence between the Earthlings' and Twin Earthlings' experiences. The Earthlings' experiences don't cause the experiences of the Twin Earthlings or vice versa, any more than one Earthling's experiences cause another's. But, in the nature of the case, the counterfactual or probabilistic dependencies between Earthling and Twin Earthling experiences are the same as those between different Earthlings' experiences, or between the experiences of different Twin Earthlings.

If, as Mill suggests, different people perceive the same thing as long as their experiences are mutually dependent in the way that different people's experiences typically are when they perceive the same thing, we'd have to say that the Earthlings and Twin Earthlings perceive the same flag.

The problem that the twin flags scenario raises for phenomenalists isn't that it prevents us from differentiating a world in which there are two indistinguishable flags from a world in which there's just one flag: we can do that by pointing out that in the two-flag world, some ideal observers would see two flags waving in tandem (one on Earth, the other on Twin Earth), whereas in a one-flag world, no ideal observer would have such a two-flag experience. The problem is that the twin-flags scenario deprives phenomenalists of Mill's account of the intersubjectivity of physical phenomena. It's one thing to explain how a twin-flags scenario differs from a single-flag scenario, another to explain how someone in the twin-flags scenario can see either flag without seeing both.

This isn't just a technical glitch in Mill's account. Kantians and metaphysical realists might seize on it as evidence that mere possibilities of sensation aren't enough for a physical world. They might argue: physical things are essentially intersubjective; the only way we can account for this feature of physical things is by supposing that different subjects' experiences of them have a common underlying source; therefore, we should identify physical things with the underlying sources of experiences, not with mere possibilities for suitably interrelated experiences.

But that's too quick. There is a way for phenomenalists to secure intersubjectivity for physical objects, construed as possibilities for experience. We can do so by locating experiences in a common ideal spacetime. Given such a spacetime, we can say that the reason why some experiences are perceptions of one flag rather than the other in the Twin Earth scenario is that the Earthlings' experiences occupy one region of ideal space, while the Twin Earthlings' experiences occupy a different region of the same space.

Our first order of business in this chapter is therefore to describe ideal worlds in which experiences occur in a common, intersubjective spacetime. To this end, I'm going to identify phenomenological relations that play the same role in ideal worlds as physical spacetime relations play in the physical world. I'll start by looking at how scientists define physical spacetime concepts, and then introduce corresponding experiential concepts through definitions modeled on the scientific ones. Parallel definitions are the best way to ensure parallel roles for the defined relations. Anyway, this is the approach I'm going to take.[3]

4.2 Physical spacetime

Suppose we have clocks that we know are accurate. (Never mind how we know they're accurate: we'll come back to that later; for now, we can imagine that God tells us they're accurate.) The exact nature of the clocks isn't important for the purposes of this illustration. All that matters is that there occurs in each one of them some repetitive observable process, like the swings of a pendulum, or the movement of a pointer around a dial, or the changes in a numerical display.

Suppose we also have probes that we know always move in straight lines at the same speed. (Never mind how we know this about the probes: we'll come back to that later; again, we can imagine that God assures us that the probes behave as described.) The exact nature of the probes isn't important. All that matters is that every probe moves in a straight line, and every probe has the same velocity as every other probe, like photons or electromagnetic waves travelling through the vacuum of empty space. Also, each probe has

[3] It's the approach that Ayer proposes, but doesn't actively pursue, in Ayer (1946: 159).

a unique identifying mark, like a serial number or barcode (similar to how waves of light can have different wavelengths).

Besides these clocks and probes, imagine that the physical contents of the universe are much as they actually are. Imagine also that there are observers located at various times and places. The job of these observers is to keep records. Every observer has one of the God-given clocks. The clocks are all synchronized, using the same method your computer uses to synchronize its clock with remote internet servers.[4] Whenever a probe enters an observer's vicinity, the observer makes a note of the time (as indicated by his clock) and records whatever other observable events take place in his vicinity at that time.

Observers periodically send records of their observations to a central institute, where specialists assign each recorded event a set of coordinates $\langle x, y, z, t \rangle$. An event's t coordinate is the time that the event's observer recorded for that event. The specialists assign x, y, and z coordinates to events in a way that's consistent with the various reported events having occurred at the various distances from one another inferrable from the observers' records. For example, if Observer A records an event e occurring near her when probe x passes through A's vicinity at t, and Observer B records an event e' occurring near him when x passes through B's vicinity at t', the specialists at the institute assign spatial coordinates to events in such a way that according to those coordinates, the distance between e and e' is the distance a probe travels in $t' - t$ amount of time (which they might express in units of probe-years, analogous to light-years).

After the specialists assign coordinates to all the reported events in the way just explained, they study the resulting coordinatization of events for signs of order and regularity. Maybe they create a computer animation based on the coordinates. In the animation, various patterns emerge: planets orbit stars in ellipses, sweeping out equal areas in equal amounts of time; massive bodies accelerate towards one another in proportion to their combined masses and in inverse proportion to the square of their distance from one another; for

[4] When Observer A's clock reads t_A, she sends a probe x towards Observer B, with a message instructing B to send a probe to A immediately upon receiving x, with an indication (from B) of what reading B's clock shows when x reaches him. When x reaches Observer B with Observer A's message, Observer B's clock reads t_B, and B sends a probe y to A, with the information A requested. When y reaches Observer A, A's clock reads t'_A. A surmises that the total time it took the probes to traverse the distance between A and B (in opposite directions) was $t_A - t'_A$, and concludes that when y reaches A, B's clock reads $t_B + (t_A - t'_A)/2$. A sets her clock to that time, and A's and B's clocks are now synchronized.

every action, there is an equal and opposite reaction; momentum is always conserved; etc. In short, when the specialists describe the physical world in terms of the coordinate system they construct based on the observers' records, they find they describe it as a world in which everything happens in accordance with a small number of intelligible mathematical rules.

There aren't observers everywhere in the universe, and there aren't clocks and probes at all times and places either. So, a lot of the world's physical events don't occur in the vicinity of any clock, probe, or observer, and these events don't get assigned any coordinates by the specialists at the institute. Still, there are facts of the matter as to what observations would have been made in the vicinity of in-fact unobserved events, had there been observers, clocks, and probes present when those events occurred. So, unobserved events also have spacetime coordinates, just undocumented ones.

Since the specialists know (from God) that the clocks and probes are reliable, they conclude that their coordinatization of events describes the true structure of physical time and space, and the true relations of the things that exist in it, and they call the simple rules that things are described as obeying when described in terms of their coordinate system "the laws of physics."

The story just told does not reflect how actual physicists map spacetime and its contents. We aren't given divinely guaranteed clocks and probes to work with. So what do we do? We take some type of seemingly repetitive process and some kind of seemingly constant motion and *assume* that the process functions like an accurate clock and the motion like a reliable probe, then assign coordinates to things on that assumption, and see what the resulting assignment looks like. If the coordinates don't reveal any pattern, we throw them out, choose some new things to treat as clocks and probes, and try again. We repeat the process until we either give up, or get a set of coordinates that describes the world in all its prima facie complexity as unfolding according to a small number of simple mathematical laws. Happily, we find ourselves in the latter scenario, due to the good efforts (and lucky guesses) of people like Kepler, Galileo, Newton, Maxwell, and Einstein. We declare the clocks and probes that yield this happy result accurate and reliable, and we declare the coordinates they deliver a true representation of the physical world's spatial and temporal features. (So, in the actual case, it's the laws of physics that play the role that God plays in the fictional case: they are what gives us confidence in the accuracy of our clocks and the reliability of our probes.)

Let's identify the basic elements of the above-described scientific attribution of spatiotemporal locations and relations to the world's physical features. There are four of them.[5]

First, there are clocks. We can think of these as physical processes in which some type of state recurs. More abstractly, we can think of clocks as a certain kind of "worldline," where a thing's worldline is just the set of all events that ever take place in the thing. Since each event occurs at a specific location in time and space, a thing's worldline traces out the thing's path through spacetime over the course of its existence. A clock's worldline is punctuated at regular intervals by a certain type of event (advance of a second-hand, nadir of a pendular motion, atomic excitation state, or whatever). Call these *chronometric worldlines.*

Second, there are probes. We can think of these as physical objects that always move at a constant speed (at least in vacuo), and that always take the shortest path between any two points they traverse (like photons); or we can think of them as effects that propagate in the same manner (like radio waves). Probes also have worldlines; call them *telemetric worldlines.*

In order to use clocks and probes to assign events spatiotemporal coordinates, clocks and probes must sometimes occur at the same time and place as one another and as the events to be assigned coordinates. In other words, there must be *intersection of worldlines.*

Finally, in order to identify certain candidate clocks and probes as accurate or constant, and the coordinate system constructed on their basis as a true representation of the world's spatiotemporal structure, we need *laws of physics.* For the coordinate system that truly represents the spatiotemporal structure of the physical world is the one that reveals physical events as unfolding according to simple, intelligible laws.

4.3 Ideal spacetime

Many experiences have phenomenal qualities that make it natural to describe them in spatial and temporal language. Just as you can describe the experience you have when you look at a stop sign as red, meaning *phenomenally*

[5] For a more detailed account of how physicists determine physical spacetime distances and locations, see the Appendix following Ch. 10.

red—red in the way that a visual experience can be red, which is different, by most accounts, from the way that a stop sign is red—you can describe the experience as octagonal, meaning *phenomenally* octagonal: octagonal in the way that a visual experience can be octagonal, which is different, by most accounts, from the way that a stop sign is octagonal. Similar remarks apply to other spatial descriptors, like size, distance, orientation, and motion.[6]

Likewise, you can describe the experience you have when you hear a balloon pop as loud, meaning phenomenally loud, and brief, meaning phenomenally brief: brief in the way that a conscious experience can be brief. Similar remarks apply to other temporal descriptors, like duration, succession, and simultaneity.[7]

Assigning spatial and temporal locations to the experiences within a single stream of consciousness is relatively straightforward: just assign coordinates to the stream's phenomenal features in a way that reflects the phenomenal distances, motions, durations, and changes that characterize the stream and its constituents. If you're looking at a Japanese flag, assign your visual experiences spatial coordinates in such a way that in terms of those coordinates, your phenomenally red experiences occupy a region that falls wholly within the region that your phenomenally white experiences occupy. If the flag appears to hang limp and then unfurl in a gust of wind, assign your flag-hanging-limp experiences earlier time coordinates than your flag-waving-in-the-wind experiences.

The challenge is to locate experiences occurring in *different* streams of consciousness, or different minds, in a common time and space. For that, we need an ideal spacetime analogous to physical spacetime.

As we saw above, the basic resources required to attribute objects and events locations in a common physical spacetime are:

- chronometic worldlines
- telemetric worldlines
- intersection of worldlines
- laws of physics

[6] Some philosophers think your experience of the stop sign is red and octagonal in the same way as the sign itself. I discuss this view (known as naive realism) in Ch. 9: see pp. 172–173.

[7] It's commonly assumed that brief experiences are brief in the same sense as brief physical events, and more generally that temporal predicates apply univocally to experiential and non-experiential phenomena. This assumption is questionable, for reasons discussed in Poincaré (1898), Russell (1926), Pelczar (2010), and Pelczar (2015: 56–83), and I don't make it in this book. (Nor do I assume that temporal predicates do not apply univocally to experiences and non-experiences: I remain neutral on the question here.)

Let's identify experiential counterparts of these resources that we can use to define an ideal spacetime.

The experiential counterparts of laws of physics are what I'll call *laws of experience*. Roughly, the laws of experience dictate that experiences occur in ways that suggest a universe governed by the laws of physics. More exactly, the world conforms to the laws of experience just in case perceptual experience tends to exhibit the sort of regularities that justify physicists in positing the laws of physics (or rather: the sort of regularities that would justify ideally intelligent and well-informed physicists in positing whatever laws of physics they would posit). The laws of physics tell physical events to obey certain rules; the laws of experience tell perceptual experiences to occur so as collectively to give the impression of physical events obeying those rules.

In our world, the physical worldlines are the theoretically observable sequences of physical events, in which each event is a cause or effect of other events in the sequence. The natural experiential counterparts of worldlines are therefore streams of consciousness.

The counterparts of chronometic wordlines are what I'll call *chronometric streams of consciousness*. These are streams that include some recurring phenomenal feature, and that facilitate an understanding of the manifold of all experiences in terms of simple laws when we use the streams the way physicists use chronometric worldlines to construct physical coordinate systems. An example is the type of stream that occurs in someone watching an accurate clock.

The counterparts of telemetric worldlines are what I'll call *telemetric streams of consciousness*. These are streams that comprise experiences as of translational motion, and that facilitate an understanding of the manifold of experiences in terms of simple laws when we use them the way physicists use telemetric worldlines in the construction of physical coordinate systems. An example is the type of stream that would occur in someone observing an inertial object moving through empty space.

Since we're going to use streams of consciousness to describe an ideal spacetime, I should say more about what streams of consciousness are, and explain how streams can intersect. The next few pages will be taken up with these details, after which we'll return to the description of ideal spacetime.[8]

[8] My main purpose in giving a detailed account of streams of consciousness is to assure readers that we can account for them without relying on physical concepts. If you don't require such assurance, you can skip ahead to p. 84.

One experience can have other experiences as parts. I'll say that an experience *X* has experiences *y* and *z* as *diachronic parts* if

1) *X* has *y* and *z* and parts,
2) *X* has no experiences as parts except for *y* and *z* (and whatever experiences might be parts of *y* and *z*),
3) *y* is an experience as of A,
4) *z* is an experience as of B,
5) neither *y* nor *z* is an experience as of A followed by B, or an experience as of B followed by A, and
6) *X* is either an experience as of A followed by B, or an experience as of B followed by A.

For example, *y* might be an experience of seeing a pendulum swing from the leftmost extreme of its arc down to its nadir, *z* an experience of seeing the pendulum rise from its nadir to the rightmost extreme of its arc, and *X* an experience of seeing the pendulum swing through a full arc from left to right; or, *y* might be a visual experience of lightning, *z* an auditory experience of thunder, and *X* an experience of lightning followed by thunder. If an experience has diachronic parts, we call it a *phenomenal process*.

Typically, one's conscious experiences occur as parts of a stream that extends from one period of dreamless sleep to another. Here I adopt what I take to be the best account of the stream of consciousness, worked out in detail by Barry Dainton: the overlap account.[9]

Two rooms overlap when they share a wall, and two stages of a relay overlap when a baton-pass is the last event of one stage and the first event of the other. Similarly, two phenomenal processes overlap just in case they have at least one, but not all, of their diachronic parts in common. Take an experience consisting of a visual impression of a flash, and an immediately following auditory impression of a bang:

Flash!-Bang!

and another experience consisting of an auditory impression of a bang, and an immediately following auditory impression of a rumble:

Bang!-Rumble

[9] Dainton (2006) presents the overlap account. As he notes, the account originates with John Foster: see Foster (1982: 258–9).

The *Flash!-Bang!* experience overlaps the *Bang!-Rumble* experience, if the *Bang!* experience that is the second experience in the *Flash!-Bang!* experience = the *Bang!* experience that is the first experience in the *Bang!-Rumble* experience. In this case, one and the same *Bang!* experience is part of two phenomenal processes: the *Flash!-Bang!* process and the *Bang!-Rumble* process.

A complete stream of consciousness is just a group of experiences that includes a given phenomenal process, together with every process that overlaps a member of the group. Typically, one phenomenal process in a stream overlaps only a small subset of other processes in the stream, but you can get from any process in the stream to any other process in it by a series of steps from one process to another overlapping process.

Suppose you watch a bowling ball roll down a lane towards some pins. Suppose you first have an experience of the ball traversing the first sixth of the lane, followed by an experience of it traversing the second sixth of the lane, and so forth, ending with an experience of the ball traversing the sixth sixth of the lane. Call these experiences e_1, e_2, e_3, e_4, e_5, and e_6.

Now suppose that e_1 and e_2 constitute a phenomenal process *A*, e_2 and e_3 a phenomenal process *B*, e_3 and e_4 a phenomenal process *C*, e_4 and e_5 a phenomenal process *D*, and e_5 and e_6 a phenomenal process *E*. So, *A* is an experience of the ball traversing a third of the lane starting from the point at which the ball starts rolling, *B* an experience of the ball traversing the third of the lane starting from a point one sixth of the way towards the pins, *C* an experience of the ball traversing the middle third of the lane, etc.

Your experiences form a stream of consciousness in this case, since your successive phenomenal processes (*A*, *B*, *C*, *D*, and *E*) overlap: *A* and *B* have experience e_2 in common, *B* and *C* have experience e_3 in common, *C* and *D* have experience e_4 in common, and *D* and *E* have experience e_5 in common. (See Fig. 4.1.)

Apart from minimal streams consisting of just two overlapping processes, every stream of consciousness has other streams as parts. The stream of consciousness I had yesterday had one part extending from morning to noon, and another from noon to the time I fell asleep last night; and there are many other ways to divide yesterday's stream into briefer sub-streams.

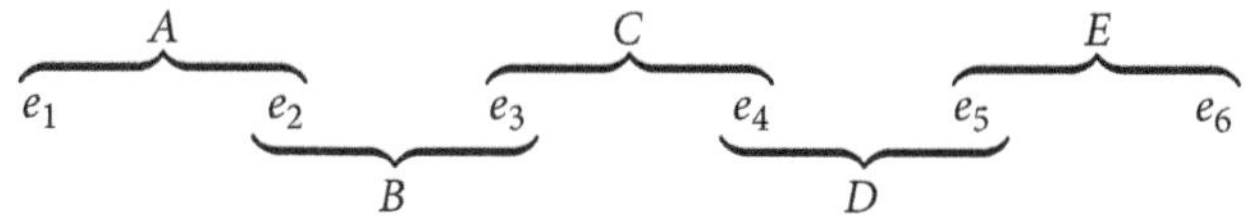

Figure 4.1 Part of a stream of consciousness

For a general definition, we can say that something is a stream of consciousness just in case it's a group of phenomenal processes such that you can get from any process in the group to any other process in the group via a series of steps from one process to another, overlapping process.[10]

A stream's successive experiences are what I'll call the *phases* of the stream. Each stream of consciousness has at least three phases (since it includes at least two processes that have at least one experience as a common part), and a typical stream has many more than three phases.

As defined, there are two kinds of stream of consciousness. What we might call "total" streams consist of overlapping phenomenal states, where an experience x is a *phenomenal state* just in case x is an experience such that there's no experience distinct from x that has all of x's parts. (Intuitively, a phenomenal state is a time-slice of the conscious mental life of some sentient being.) A different kind of stream is a longitudinal section of a total stream. For example, if you watch a robin hopping around your yard, your total stream, which consists of experiences of grass, trees, sky, birds, etc., also includes a sequence of overlapping experiences of the robin. This sequence, which is as it were a thread or current running through your total stream, is itself a stream of consciousness, as I'm using this term.

Just as intersecting physical worldlines are worldlines that have a physical event in common (such as a collision of two objects, or a lightwave reflecting off a mirror, or two photons simultaneously entering the same eye), intersecting streams of consciousness are streams that have an experience in common.

There are various ways this can happen.

One is by being two parts of the same ordinary stream of consciousness. For example, experience e_3 in Fig. 4.1 is the last phase of the stream consisting of experiences e_1, e_2, and e_3, and the first phase of the stream consisting of experiences e_3, e_4, e_5, and e_6.

Another way two streams can have an experience in common is by splitting off from a single stream, as in the leftmost diagram in Fig. 4.2.[11] One stream consists of auditory experiences *Do*, *Re*, *Mi*, *Fa*, and *So*, where the *Do* and *Re* experiences belong to an experience as of Do followed by Re,

[10] If we want to make this more precise, we can start by defining the ancestral of the overlap relation as the binary relation $\mathcal{O}$ such that for any experience x and any experience y, $\mathcal{O}(x, y)$ just in case x overlaps y, or there's an experience z such that x overlaps z and z overlaps y, or there's an experience z and an experience z' such that x overlaps z and z overlaps z' and z' overlaps y, or . . . etc. Now we define a stream of consciousness as any group of phenomenal processes such that every member of the group bears $\mathcal{O}$ to every other member of the group.

[11] Figure 4.2 is to be read from bottom to top: you can think of the experiences lower down in the diagram as occurring earlier than the ones higher up.

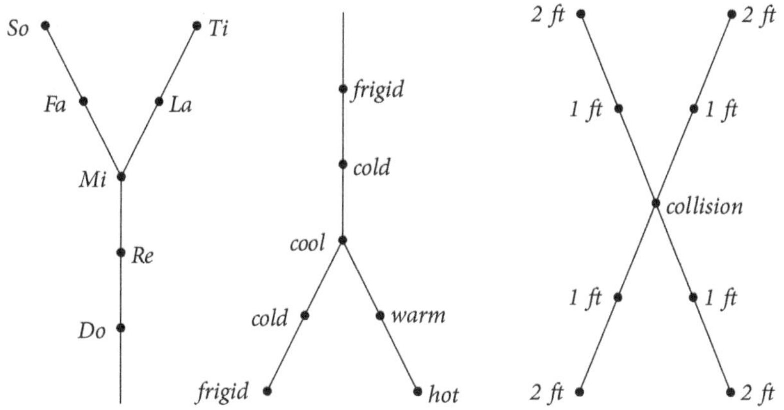

Figure 4.2 Splitting, merging, and intersecting streams

and similarly for *Re* and *Mi*, *Mi* and *Fa*, and *Fa* and *So*. The other stream consists of auditory experiences *Do*, *Re*, *Mi*, *La*, and *Ti*, where the *Do* and *Re* experiences belong to an experience as of Do followed by Re, and likewise for *Re* and *Mi*, *Mi* and *La*, and *La* and *Ti*. The *Do* and *Re* experiences (as well as the experience as of Do then Re that they constitute) occur in both streams.

Which of the streams that branch off from the *Mi* experience, if either, is a continuation of the mental life in which the *Do* and *Re* experiences occur? This is a challenging question that need not detain us here. The important thing is that the described scenario is perfectly coherent: from a first-person standpoint, it's indistinguishable from a scenario in which one person listens to the note sequence *Do-Re-Mi-Fa-So*, and another person listens to the sequence *Do-Re-Mi-La-Ti*.[12]

Two streams can also have an experience in common by merging into a single stream. The central diagram in Fig. 4.2 depicts one stream that starts with (1) an experience of the surrounding air changing from frigid to cold, continues with (2) an experience of change from cold to cool, and then continues further with (3) an experience of a change from cool back to cold, and finally (4) an experience from cold to frigid. The figure also

[12] The *Do*, *So*, and *Ti* experiences can't all occur in the same mind, since the *So* and *Ti* experiences clearly don't. One possibility is that the person in whom the *Do* and *Re* experiences occur partially survives as someone who hears the initial sequence continue *Mi-Fa-So*, and partially survives as someone who hears the sequence continue *Mi-La-Ti*: see Parfit (1984: 298–302).

depicts a stream that begins with (a) an experience of change from hot to warm, continues with (b) an experience of change from warm to cool, then (c) an experience of change from cool to cold, and finally (d) an experience of change from cold to frigid. These are all assumed to be diachronically complex experiences consisting of two experiences each. The *cold* part of (1) is the same as the *cold* part of (2), the *warm* part of (a) is the same as the *warm* part of (b), and the *cool* part of (2) is the same as the *cool* part of (b). The *cool* experience is also part of the cool-to-cold experience, whose *cold* part is the same as the *cold* part of the final cold-to-frigid experience.

In this scenario, the stream beginning with a frigid experience merges with the stream beginning with a hot experience, continuing from there as a single stream of consciousness in which it feels like there's a steady drop in temperature. If Beth has the stream extending from *frigid* to *cool*, and Bill the stream extending from *hot* to *cool*, it's a challenge to say which, if either, of the two has the experiences in the stream extending from *cool* to *frigid*. The important point is that the described scenario is perfectly coherent: from a first-person standpoint, it's the same as if one person felt the temperature rise then fall, and another person felt the temperature steadily fall. Depending on the further details of the case, the post-merger experiences might include some confusing memory experiences, but these would only make the case peculiar, not impossible.

A final way for two streams to have a phase in common is by crossing or intersecting, i.e. at once merging and splitting, as in the rightmost diagram in Fig. 4.2. Suppose Efren and Earl see two billiard balls roll towards one another, collide, and rebound. Efren has a stream of consciousness consisting of (1) an experience of the distance between the balls decreasing from two feet to one foot, which starts with a *balls two feet apart* experience and ends with a *balls one foot apart* experience, (2) an experience of the distance decreasing from one foot to zero feet, which starts with a *balls one foot apart* experience and ends with a *balls colliding* experience, (3) an experience of the distance between the balls increasing from zero feet to one foot, which starts with a *balls colliding* experience and ends with a *balls one foot apart* experience, and (4) an experience of the distance increasing from one foot to two feet, which starts with a *balls one foot apart* experience and ends with a *balls two feet apart* experience. Experiences (1), (2), (3), and (4) overlap in the usual way. Earl has a stream of consciousness exactly like Efren's.[13]

[13] Here it seems reasonable to think that both Efren and Earl persist throughout the scenario, though it might be hard to know which post-collision experiences occur in whom. The

In this case, as in the others, there's an intersection of total streams, and also an intersection of substreams. Suppose the balls in question are the blue 2 ball and the red 3 ball. Then there's a stream that starts with Efren's experience of the blue ball rolling across the felt until it collides with the red ball and then continues on its way, as well as a stream that starts with Earl's experience of the red ball rolling across the felt until it collides with the blue ball and then continues on its way, where the experience of collision that occurs in Efren's stream is the same as that which occurs in Earl's.

Above, we introduced laws of experience as the experiential counterparts of laws of physics, one of the four physical resources required to define physical spacetime. Streams of consciousness give us experiential counterparts of the three remaining resources: chronometric worldlines, telemetric worldlines, and intersecting worldlines.

The experiential counterparts of chronometric worldlines are streams of consciousness having the sort of phenomenal features that characterize our actual streams of consciousness when we perceive reliable clocks in operation: what I've called chronometric streams. They're streams in which some phenomenal property recurs, and that, when used in combination with telemetric streams to assign coordinates to experiences, yield a sense-making map of the experiential manifold.

The experiential counterparts of telemetric worldlines are streams of consciousness having the sort of phenomenal features that characterize our actual streams of consciousness when we perceive uniform linear motion: what I've called telemetric streams. They're streams in which some object appears to move (or some effect appears to propagate) through space, and that, when used in combination with chronometric streams to assign coordinates to experiences, yield a sense-making map of the experiential manifold.

The experiential counterparts of intersecting physical worldlines are intersecting streams of consciousness, understood as just explained. For example, there could be a stream of visual experiences of a probe that moves from an observer A to an observer B, and another stream of experiences of a probe that moves from B to A, where the last experience in the first stream is the first experience in the second; or, there could be a stream of experiences of a probe that moves from A past B to C, and a stream of experiences of a probe

important point is that the scenario is perfectly intelligible: from a first-person point of view, it's the same as if Efren's and Earl's streams had flowed along in the normal way, without having any of their constituent experiences in common.

that moves from D past B to E, where the two streams have their "probe located near B" experience in common.

Earlier, I described a scenario in which observers using synchronized clocks record when various probes pass by them, and what other events take place in their vicinities at those times, periodically sending copies of their records to a central institute, where specialists use them to construct a system of spacetime coordinates for physical events. If we imagine that there are enough observers that each probe is always observed by at least one observer, and stipulate that some observers' streams intersect, the scenario contains all the experiential elements required to assign various experiences in it locations in ideal spacetime.

Just as specialists at the scientific institute described earlier can assign coordinates to physical events based on the records that the observers send them, they can assign coordinates to phenomenal events, based on the experiences that are the basis of those records. If the coordinates portray the observers' experiences as unfolding according to simple intelligible laws, the specialists conclude that the coordinates accurately represent the locations of and relations among the observers' experiences in ideal spacetime.[14]

There's nothing mysterious about ideal spacetime: it's just a framework for describing experiences. That such a framework exists and is useful is surprising, but only for the same reason it's surprising that there's a useful system of spacetime coordinates for describing physical events. Physical events occupy locations in a common physical spacetime by occurring (or being apt to occur) in worldlines with unmysterious features, relating in unmysterious ways; experiences occupy locations in a common ideal spacetime by occurring (or being apt to occur) in streams of consciousness with equally unmysterious features, relating in equally unmysterious ways.

The only asymmetry between the physical and ideal frameworks is that intersection gets defined in the ideal framework, but not in the physical framework. It's actually unclear how we're supposed to understand worldline intersection in the physical case. It's tempting to say that intersection of physical worldlines is something like direct interaction. But unless "direct" just means "at zero distance in time or space," there's no obvious reason why

[14] As in the physical case, experiences can have locations and relations in ideal spacetime without there actually being any chronometric or telemetric streams of consciousness: it's enough if various such streams would intersect with each other and other streams in the right ways, if there were a sufficient abundance of chronometric and telemetric streams.

one event couldn't directly cause another event that occurred far away from it in time or space.

What mappers of physical events actually do is provisionally assume that a given event belongs to multiple causal chains of events and, if the assumption facilitates a useful mapping of events, declare that those event-chains intersect at the given event. This is in contrast to the ideal case, where we have defined intersection as partial identity (two streams of consciousness intersect if and only if they have an experience in common). If you don't like that definition, you can follow the physicists' example, and declare that streams intersect wherever it's useful (for map-building purposes) to assign their phases the same ideal coordinates.[15] But it's an advantage of the approach taken here that it gives us a direct, independent grasp of what it is for two streams of consciousness to have a phase in common.

In us, chronometric and especially telemetric streams of consciousness typically consist of visual experiences, like experiences as of the motion of a clock's second-hand or propagating wave. More generally, visual experience is our main source of information about how things are arranged in time and space.

Our main source, but not our only source: we also have non-visual experiences by which we perceive things as having spatial locations and moving or changing over time. For example, tactile experience, or a combination of tactile and proprioceptive experience, can inform us of the locations of various objects within our immediate vicinity, as when groping one's way through a dark room, or feeling something crawling up one's leg. Tactile experience often combines with visual experience to form a multi-modal conscious appearance of something located in space, as when you touch and see an object at the same time.

In typical human experience, other sensory modalities make relatively minor contributions to spatial phenomenology, though auditory experience often tells us the direction from which a sound comes, and, in combination with visual or tactile experience, can inform us of the location of a sound's origin. Our senses of taste and smell contribute very little to our spatial phenomenology, though when combined with experience from other modalities, we can assign flavors and odors locations in space, as when smelling a rose that you hold up to your nose, or tasting a peppermint in

[15] Broadly speaking, this is the approach taken by Foster (1982).

your mouth. In an ideal spacetime, all these experiences belong to streams of consciousness that allow us to assign them ideal spacetime locations.

4.4 Ideal events

Now that we've defined ideal spacetime, we can define ideal events as certain groups of experiences co-located in ideal spacetime, and an ideal world as the largest set of ideal events occurring in the same ideal spacetime as a given ideal event. This will answer the main question of this chapter, of what phenomenalists take physical reality to be a possibility for. Physical events are possibilities for ideal events; the physical world is the possibility for an ideal world.

First, we define an *eventlike* group of experiences as a set of experiences occurring at the same location in ideal spacetime, but in different minds, and linked by relations of conditional probability:

> Some experiences are an **eventlike group** if and only if (1) the experiences occur in different streams of consciousness, (2) the experiences are co-located (in ideal spacetime), and (3) each experience x in the group is such that for every other experience y in the group, $\Pr(x|y) > \Pr(x)$.

Experiences belonging to different eventlike groups can occur at different locations in the same ideal spacetime. Given one eventlike group of experiences, the totality of all eventlike groups existing in the same ideal spacetime as that group is what I'll call an *ideal manifold* of experiences:

> An **ideal manifold** is a totality T of experiences such that (1) each experience in T belongs to an eventlike group of experiences, (2) each experience in T occurs in the same ideal spacetime as every other experience in T, and, (3) T includes every experience that belongs to an eventlike group and occurs in the same ideal spacetime as an experience in T.

Identifying physical events with possibilities for eventlike groups of experiences, and physical worlds with ideal manifolds, would solve the problem raised at the beginning of this chapter, of how to reconcile phenomenalism with the intersubjectivity of the physical. Suppose, for the sake of illustration, that there really is a Twin Earth in our universe. Now consider an ideal counterpart of our physical world. In this ideal world, there are experiences of *Titanic* striking an iceberg, and experiences of Twin *Titanic* striking an indistinguishable iceberg. Taken all together, these experiences are simultaneous and interdependent (in the same way as the experiences in each group considered separately); the experiences in each group also resemble those in the other just as they resemble one another. However, the experiences in the first group occupy a different region of ideal space from those in the second group. This is what makes the experiences in the two groups experiences of different (though qualitatively indistinguishable) maritime disasters.

However, it would be a mistake for phenomenalists to identify all possibilities for eventlike groups of experiences with physical events, as the following examples show.

If two people view Fig. 4.3 in the same copy of this book at the same time, their experiences (or some of them) are co-located in ideal spacetime and interdependent in the usual statistical or counterfactual sense (if the book were rotated, the two people's experiences would change in parallel ways). Still, there's no physical thing on the page whose A zone is darker than its B zone (the color difference is an illusion). So, although the people's experiences of the figure constitute an eventlike group, the possibility for those experiences is not sufficient for the existence of a physical thing whose A zone is darker than its B zone.

If two people come across the footprints pictured in Fig. 4.4, the footprints might look to both of them like foot-shaped reliefs rising above the surrounding sand. If a beetle crawls across one of the prints, it may look to both of them like the beetle is climbing up, over, and down a foot-shaped hill, rather than into, across, and out of a foot-shaped hole. So the experiences are interdependent. They're also co-located in ideal spacetime. Yet the reality is

Figure 4.3 Checker Shadow Illusion
Wikimedia / Pbroks13Wikimedia / Pbroks13

Figure 4.4 Footprints
Wikimedia / Janhannes (CC BY-SA 4.0)

Figure 4.5 Floor Tile Illusion
Courtesy of Casa Ceramica

that there is a hole, not a hill. So, although the people's experiences of the prints constitute an eventlike group, the possibility for those experiences is not sufficient for the existence of a foot-shaped mound of sand.

If two people look at the floor pictured in Fig. 4.5 from the depicted angle, it will looked warped to both of them. Their experiences are interdependent (if the light dims, both people will see the hall darken), and they occupy the same region of ideal spacetime. But there is no warp—the floor is level—so the possibility for this eventlike group of experiences isn't sufficient for the existence of a warped floor.

As these examples show, illusions provide many examples of eventlike groups of experiences whose possibility does not entail the existence of a corresponding physical event. We can imagine more extreme cases. If two people are plugged into a virtual reality machine that gives them the same fully realistic perceptual experiences at the same time (maybe as part of a pre-programmed virtual-reality field trip to the Moon), their experiences are interdependent and, perhaps, ideally co-located, but they're not manifestations of real lunar craters, mountains, etc. Likewise, we can imagine a society of people with "reality-augmenting" neural implants that induce fully realistic coordinated group hallucinations of Pokémon characters (or whatever). Such people's experiences aren't perceptions of actual physical things, even if their experiences constitute eventlike groups.

All of the experiences described above are phenomenally indistinguishable from possible veridical experiences, but they differ from veridical experiences in how they relate to other possible experiences. There are possible experiences that would reveal that areas A and B of the diagram in Fig. 4.3 are the same color; for example, if we had experiences of cutting the A and B areas out of the page and putting them next to each other, we'd have experiences of the cuttings having the same color. There are possible experiences that would reveal the things we see at the beach as holes rather than mounds, such as the experiences we'd had if we got down and looked at them from a lower angle, or reached down and touched them. There are possible experiences that would reveal the trick hallway to be flat rather than warped, like the experiences you'd have if you walked down the hall. There are possible experiences that would reveal VR or augmented-reality experiences as such, like the experiences of the people running the VR facility, or the experiences of the surgeons who install the neural implants, or the experiences people have when volunteering to enter the VR pod or receive the neural implants.

In short, the experiences described above, though intrinsically indistinguishable from possible veridical experiences, don't relate to other experiences the way we normally expect veridical perceptions of physical things to relate to each other. What way is that, exactly?

This is a surprisingly hard question to answer. I'll return to it in the next section. For now, the best answer I can give is that the "right way" is the way that your current experiences relate to the other experiences you've had, and that the experiences you've had in what you take to have been dreams, hallucinations, and illusory experiences do *not* relate to the rest of your experiences. I'll call this way that some but not all experiences have of relating to other experiences *phenomenal coherence*, or "coherence," for short.

An ideal manifold might contain some eventlike groups whose constituent experiences do, and other eventlike groups whose constituent experiences do not, cohere with the rest of the manifold. The eventlike groups whose constituent experiences do cohere with the rest of the ideal manifold in which they occur are what I'll call *ideal events:*

> An **ideal event** is an eventlike group of experiences comprising experiences that cohere with the rest of the ideal manifold in which the experiences occur.

In other words, an ideal event is a group *G* of experiences that (1) don't all occur in the same stream of consciousness, (2) are interdependent, (3) have the same location in ideal spacetime, and (4) cohere with the rest of the ideal manifold that *G*'s experiences belong to.

An ideal world is the totality of all ideal events that occur in the same ideal spacetime:

> An **ideal world** is a totality of ideal events that includes all and only the ideal events whose experiences occur in the same ideal spacetime as the experiences constituting a given ideal event.

Intuitively, an ideal world is a state of affairs wherein experiences occur in groups like those that occur when people simultaneously perceive the same physical thing, and in which there are streams of consciousness connecting such groups. The pattern of intersections among such streams is the basis for assigning various experiences locations in ideal spacetime, just as the pattern of intersections among physical worldlines is our basis for assigning physical events locations in physical spacetime.

Just as physical objects and processes are suitable combinations of physical events, ideal objects and processes are suitable combinations of ideal events.

Physical processes are sequences of causally related physical events. Ideal processes are sequences of similarly related ideal events. Two ideal events belong to the same ideal process just in case each of the events' constituent experiences occurs in the same stream of consciousness as one of the other event's constituent experiences. You can think of an ideal process as a bundle of streams of consciousness each cross-section of which is an ideal event.

The ideal counterparts of ordinary physical objects are ideal processes that satisfy certain criteria. An ideal tree, for example, consists of an ideal process such as one would expect to occur in a scenario where ideal observers perceive a tree over some period of time: streams of consciousness featuring experiences of germination, growth, fruiting, etc.

"Certain criteria" is deliberately vague. The details depend on our best reconstruction of folk ontology. Such a reconstruction is beyond the scope of this study, as is the question whether there is actually any good reconstruction of folk ontology (some metaphysicians think there isn't, at least when it comes to folk ontology of the physical world).[16]

My own view, which I won't defend here, is that a tolerably good reconstruction of folk ontology is possible along the lines that Amie Thomasson recommends. Roughly, Thomasson identifies ordinary physical objects with the satisfiers of the commonly-accepted application conditions for the sortal terms we use in everyday talk about physical things ("rock," "chair," "tree," etc.). Since the main thing that informs everyday talk about physical things is everyday experience, Thomasson's account fits well with phenomenalism.[17]

Thomasson isn't committed to phenomenalism, of course, and neither are phenomenalists committed to Thomasson's account of ordinary objects. Phenomenalists can accept any reconstruction of folk ontology, provided that, like Thomasson's, it maintains a robust connection between the concept of a physical object and that of a possibility for experiences of a physical object. Since it's hard to see how any plausible reconstruction of folk ontology could fail to maintain such a connection, the job of reconstructing the concept of an ordinary object is one that phenomenalists can safely leave to others.[18]

4.5 Reduction or elimination?

"Physical things are nothing but possibilities of sensation" can be taken two ways. Taken one way, it means that the physical facts about our world reduce to facts about possibilities of sensation; this is a kind of reductionism about

[16] See, e.g., Merricks (2001). [17] See Thomasson (2007: 176–87).

[18] Phenomenalism is also compatible with the wholesale rejection of folk physical ontology. However, like Thomasson, I'm not persuaded by the arguments for rejecting it.

the physical. Taken another way, it means that there really are no physical things, but only possibilities of sensation; this is a kind of eliminativism or antirealism about the physical.[19]

I advocate reductive rather than eliminative phenomenalism. However, in our present state of knowledge, I don't think we can completely rule out the eliminative possibility.

Recall that a defining characteristic of an ideal event is that its constituent experiences relate coherently to other experiences occurring in the same ideal spacetime. Whether we should prefer reductive phenomenalism to eliminative phenomenalism or vice versa depends on the nature of this coherence.

We're all confident that we can distinguish our veridical experiences from our non-veridical ones with a high rate of success, but, as G.E. Moore points out, it's surprisingly hard to say exactly how we accomplish this:

> I have, no doubt, conclusive reasons for asserting that I am not now dreaming; I have conclusive evidence that I am awake: but that is a very different thing from being able to prove it. I could not tell you what all my evidence is; and I should require to do this at least, in order to give you a proof. (Moore 1939: 149)

If I want to figure out whether one of my experiences is a veridical waking experience or a dream experience, all I can do, ultimately, is consider how the experience relates to the other experiences that I'm having or can remember having. I consider whether the experience coheres with my other experiences, and judge the experience veridical or not based on that. But what does this "considering whether the experience coheres with my other experiences" amount to?

There are two possibilities to consider.

One is that our inclination to count some experiences as dreams and others as non-dreams reliably tracks a relationship that the non-dreams bear to other experiences, but the dreams do not. Our inclination might have this tracking function, even if we can't give a detailed description of the relationship (among experiences) that the inclination tracks. (Similarly, our

[19] There's a similar ambiguity in Berkeley's idealism: sometimes he seems to advocate reductive idealism (e.g., Berkeley 1710/1901: §§1, 99), but sometimes he seems to advocate eliminative idealism (e.g., Berkeley 1710/1901: §20). Most scholars interpret Berkeley as a reductionist, which is how I interpreted him in Ch. 2.

inclination to classify some foods as sour and others as non-sour reliably tracks a chemical feature that sour foods have but non-sour foods don't, even if we can't describe that feature in detail.)

If our inclination performs this tracking function, there is an objective fact of the matter about whether any given experience coheres with other possible experiences, even if for some reason we're not able to specify the coherence relation very exactly. In this case, phenomenalism is best understood as a reductive theory that analyzes physical things partly in terms of coherence among possible experiences.

The other possibility to consider is that our inclination to judge certain experiences as cohering with our other experiences isn't based on a recognitional capacity at all, but is a kind of non-classificatory response behavior, like the disgust reaction.

Disgust is part of a behavioral immune system that evolved to prevent us from coming into contact with infectious pathogens. Since getting infected often has dire consequences, it's best to be disgusted by a lot of harmless things if that's what it takes to be disgusted by all the infectious things. This, together with the wide variety of things that actually are infectious, makes it unsurprising that the things that disgust us don't all have some single quality that makes us judge each of them disgusting. Moldy food disgusts us because it has one feature, which differs from the disgust-inducing feature of feces, which differs from the disgust-inducing features of cockroaches and leeches.

Maybe our sense that certain experiences cohere with other experiences is a similarly non-classificatory response to sensory inputs. We take a different psychological posture towards experiences we consider veridical from what we take towards experiences we consider non-veridical. Well, just as different factors can elicit the disgust reaction in us, it might be that different factors can elicit a "realistic" psychological posture towards a given experience. Maybe we've evolved a tendency to take this posture towards our experiences by default; at least, treating our experiences as veridical until proven otherwise seems more advantageous than the opposite policy (the one Descartes experiments with in his *Meditations*).

If this is right, it's not surprising if some experiences strike us as coherent due to relating to other experiences one way, while other experiences strike us as coherent due to relating to other experiences in other ways. Just as there's no single quality, "disgustingness," that makes us judge things disgusting, but a mixed bag of qualities whose only interesting commonality is that they're the things that elicit a disgust reaction from us, it may be that there's no single relation, "coherence," that makes us judge experiences veridical, but only a mixed bag of relations to other experiences whose only interesting commonality is that they're the relations that elicit a realistic psychological posture from us. If so, phenomenalism is best understood as an eliminative theory of the physical, in which talk about possibilities of sensation replaces talk about physical things, instead of physical things being reduced to possibilities of sensation.

Like I said, I'm defending a reductive form of phenomenalism in this book. Here I add the caveat that if phenomenal coherence turns out not to be an objective relationship among experiences, but only a non-classificatory way we have of reacting to various experiences, we should switch to eliminative phenomenalism.

5
What Kind of Possibility?

An ideal world is a network of experiences satisfying the criteria described in the preceding chapter. According to traditional idealists, like Berkeley and Leibniz, the physical world we inhabit *is* an ideal world. According to phenomenalists, the physical world is the *possibility* for an ideal world that does not, as far as we know, actually exist.

The question we need to answer now is: what does "possibility" mean in this context? Among all the *logically* possible ideal worlds, which one is possible in the way that makes its possibility equivalent to the actual existence of the physical world we inhabit?

My answer, basically, is that the logically possible world for which there's a possibility in the relevant sense is the one that our own actual experiences would belong to, if they belonged to an ideal world. In this chapter I develop this answer in a way that avoids the objections primarily responsible for phenomenalism's historic decline.

In §5.1, I bring the traditional phenomenalist account of possibilities of sensation up to date, by interpreting sensation conditionals according to the now-standard analysis of counterfactuals, which wasn't available at the time phenomenalism came under attack 70 years ago. According to this analysis, saying that a certain ideal world is the one our experiences would belong to if they belonged to an ideal world is the same as saying that a certain ideal world is the one our experiences do belong to in the logically possible worlds most similar to the actual world, among those in which our experiences belong to an ideal world.

In order for phenomenalists to make use of this analysis, there must be some non-physical respect in which a possible world can be more or less similar to the actual world. In §5.2, I argue that the required dimension of similarity is similarity with respect to phenomenal probabilities: conditional probabilities related to possible experiential states of affairs.

Section 5.3 summarizes the phenomenalist theory I favor.

Phenomenalism: A Metaphysics of Chance and Experience. Michael Pelczar, Oxford University Press.
 DOI: 10.1093/oso/9780192868732.003.0005

5.1 Counterfactual possibility

It will be useful to have a word for the kind of possibility that characterizes experiential structures that are possible in the sense relevant to phenomenalism, as opposed to mere logical or metaphysical possibility. I'll call it *counterfactual possibility*.

Counterfactual possibility is supposed to be a modality situated somewhere between actual existence and mere metaphysical possibility. Counterfactual possibilities are states of affairs that are waiting in the modal wings, so to speak.

What we need to do now is replace these metaphors with a precise definition. Here's how I'm going to go about this. First I'm going to tell you what it is for an ideal world to be counterfactually possible. Then I'll define counterfactually possible sensations, streams of consciousness, etc. as those that occur in the ideal world for which there is a counterfactual possibility.

Here's my definition of "counterfactually possible ideal world":

> For all logically possible ideal worlds x, there is a counterfactual possibility for x just in case x is the ideal world that our own actual experiences are most likely to belong to in the metaphysically possible worlds most similar to ours, among those in which our actual experiences belong to an ideal world.

If we follow the usual practice of defining the "nearest ϕ worlds" as the metaphysically possible worlds most similar to ours among the metaphysically possible worlds in which ϕ, we can put this more succinctly as follows:

> The counterfactually possible ideal world is the ideal world that our own actual experiences most likely belong to in the nearest worlds in which our actual experiences belong to an ideal world.

The repeated occurrence of the word "world" in the statement above is distracting, so, for readability's sake, let's use "ideal network" as a synonym for "ideal world." Then we can restate the above as:

> The counterfactually possible ideal network is the one our own actual experiences most likely belong to in the nearest worlds in which our actual experiences belong to an ideal network.

Since I'll have occasion to refer back to this definition repeatedly in what follows, let's give it a name: "the possibility postulate."

My basic position is that the physical world consists of the probabilistic facts about experience—"phenomenal probabilities," as I call them—in virtue of which a certain ideal world is the one for which there's a counterfactual possibility. But before going any farther, there are three questions I need to answer about the possibility postulate. First, what do I take as the measure of similarity among possible worlds? Second, why the qualification "most likely"? Third, why the reference to our actual experiences?

Those familiar with the analysis of counterfactual conditionals due to Robert Stalnaker and David Lewis will recognize the possibility postulate as equivalent, by that analysis, with the claim that the counterfactually possible ideal world is the one that our own actual experiences would most likely belong to, if they belonged to an ideal world.[1] One nice feature of the Stalnaker–Lewis analysis is that it pre-empts Chisholm's argument from perceptual relativity. Chisholm's argument was that no non-trivial statement of the form,

> If experiential state of affairs ϕ existed, then experiential state of affairs ψ would exist.

is true, since for any logically independent experiential states of affairs ϕ and ψ, ϕ might exist in the absence of ψ due to unusual physical circumstances (involving hallucinogens, optical illusions, or whatever).

The Stalnaker–Lewis analysis of counterfactual conditionals stops Chisholm's objection in its tracks. I seem to see a cantaloupe in front of me, and I believe that if I were to have reaching-out-to-touch-the-melon experiences, I'd have feeling-a-cool-rough-surface experiences. Is my belief mistaken? Maybe; but it's not shown to be mistaken by the mere fact that there's a possible scenario in which what I take to be a cantaloupe is an optical illusion. That fact doesn't imply that it's false that I have the described cool-and-rough tactile experiences in all of the worlds most similar to the actual world among those in which I have the reaching-out-towards-a-melon experience.

[1] For the standard analysis of counterfactual conditionals, see Stalnaker (1968) and Lewis (1973b). Stalnaker's and Lewis's analyses are very similar, but I'll be assuming Lewis's, which is slightly more conservative than Stalnaker's.

This is good news for phenomenalists, but we still need to hear what makes any given possible world more or less similar to the actual world.

There's one answer that isn't available to phenomenalists: in judging the nearness of various possible worlds to the actual world, phenomenalists can't appeal to physical similarities or dissimilarities between worlds, since in the phenomenalist view, a world's physical features depend on how near the world is to the actual world, in some yet-to-be-specified respect. Since phenomenalism is an attempt to reduce the physical facts to facts about possibilities for experience, phenomenalists can't turn around and reduce facts about possibilities for experience to facts about how much our world physically resembles other possible worlds.

Assuming that materialism isn't true, phenomenalists can appeal to worlds' occurrent phenomenal features, but that won't get us very far. Among possible worlds in which our actual experiences belong to ideal networks, some are worlds in which our experiences belong to one ideal network, and some are possible worlds in which our experiences belong to very different ideal networks. (You can think of these different possible networks as different ways our existing stock of experience might grow as we explore more and more of our world, depending on the physical features of the parts of the world we have yet to explore.) Similarity in occurrent phenomenal respects plays some role in deciding which possible worlds are nearest our own, but it's at most a small role, and certainly not enough to single out a unique ideal world as the one for which there's a counterfactual possibility.

If the measure of similarity among worlds isn't physical similarity, and if similarity in occurrent phenomenal respects is inadequate as a measure of modal proximity for phenomenalists' purposes, what's left?

What's left, I suggest, are phenomenal probabilities: chancy facts about experience. The worlds closest to ours (among those where our actual experiences belong to an ideal network) are the ones most similar to ours in their phenomenal probabilities (among those where our actual experiences belong to an ideal network). The relevant similarity among possible worlds is similarity with respect to phenomenal probabilities.

5.2 Phenomenal probabilities

Phenomenal probabilities are objective conditional probabilities related to experience. By "conditional probabilities," I mean facts of the form:

Pr (state of affairs *P* exists | state of affairs *Q* exists) = x

E.g. the chance of drawing a black card from a deck given that it's a standard deck = $\frac{1}{2}$

or:

$x <$ Pr (state of affairs *P* exists | state of affairs *Q* exists) $< y$

E.g. the chance of winning the national lottery given that you buy a ticket is between 0 and $\frac{1}{1{,}000{,}000}$

or:

Pr(state of affairs *P* exists | state of affairs *Q* exists) > Pr(state of affairs *P* exists)

E.g. the probability of getting lung cancer given that you smoke is greater than the unconditional probability of getting lung cancer

—etc. By calling a conditional probability "objective," I mean that it's not just a measure of the ignorance of beings (like us) with limited information and intelligence, but a feature our world has independent of any cognitive agent, in the same way it has various causal and spatiotemporal features.

I assume without argument that there are objective conditional probabilities. This assumption, though not uncontroversial, is widely held, including by such arch-realists as Karl Popper and David Lewis.[2]

Following Popper and others, I also assume that there are objective conditional probabilities—"chances," for short—related to non-existent states of affairs. In particular, I assume that we may speak of the chance that *q* given that *p* in cases where the objective probability that *p* is zero or infinitesimal. Though also not entirely uncontroversial, the latter assumption receives support both from intuition and from the existence of consistent axiom systems satisfying the formal criteria generally thought to encapsulate our ordinary concept of probability and in which formulae of the form, "Pr(p|q)" ("the chance of p given that q") are defined even when Pr(q) is zero or infinitesimal.

As for intuition: intuitively, the chances that I'd have won the lottery given that I'd bought a ticket are lower than the chances that I'd have lost given that

[2] See Popper (1959/1992: 137; 1983: 281–300) and Lewis (1981: 263).

I'd bought a ticket, even if I didn't buy a ticket; intuitively, the likelihood that the *Venus de Milo* falls given that it's heaved off the observation deck of the Empire State Building is greater than the likelihood that the *Venus* hangs suspended in air given that it's heaved off the observation deck, even if the *Venus* never gets heaved off the observation deck; intuitively, the probability of finding half as many radon atoms in our vacuum chamber given that we look inside it four days after filling it with newly-created radon atoms is greater than the probability of finding one tenth as many radon atoms in our chamber given that we look inside it four days after filling it with newly-created radon atoms, even if we never put any radon atoms in the chamber.

As for the axiom systems, there are various that accommodate probabilities conditional on states or events with zero or infinitesimal likelihood, while arguably capturing our pre-theoretical notions of what probability is. These include systems defined in terms of Popper functions, as well as various systems defined in terms of non-Archimedean probability functions which can take infinitesimal values (in addition to the usual unit-interval real values).[3]

When we think about applications of probability, we usually think of physical applications—coins, lotteries, gases, radioactive decay—but there's nothing to stop us from applying probability to the experiential domain. We have as much reason to think that our world includes chancy facts related to experience as to think that it includes chancy physical facts; this is true even if facts about experience aren't physical facts. What I'm calling phenomenal probabilities are examples of such facts: they're just objective probabilities like those listed at the beginning of this section, with P and Q standing in for propositions describing purely experiential states of affairs.

For example, in our world, the probability that something is an experience as of a statue of a robed lady, given that it's an experience as of a large, blue-green neoclassical statue of a crowned, torch-bearing lady standing on

[3] For the first approach, see, e.g., Popper (1959/1992: 329–55), Stalnaker (1970), and Roeper and Leblanc (1999: 10–16); for the second, see, e.g., Blume et al. (1991), Hammond (1994), and Benci et al. (2013). McGee (1994), Halpern (2010), and Brickhill and Horsten (2018) discuss the relationships among these systems. Hannes Leitgeb argues for a general analysis of counterfactual conditionals in terms of Popper functions. I'm sympathetic to Leitgeb's proposal, but for present purposes it's more than we need. Again, what we're trying to articulate here is the idea that some metaphysical possibilities are closer to being actual than others; for this, it's enough to show that some metaphysically possible worlds are closer than others to the actual world, when we take similarity with respect to chancy facts as our measure of proximity. Whether we can use probabilistic concepts to capture the meaning of counterfactual conditionals in everyday language is an interesting but separate question. For Leitgeb's analysis, see Leitgeb (2012a; 2012b), and for critical discussion, Hájek (2014b: 243–9).

a pedestal on a small island at the mouth of a river lined with skyscrapers, is high: higher than in many other possible worlds. Also high is the probability that something is a stream of consciousness culminating in an experience as of a dusty barren plain, given that it's a stream with the phenomenal features that characterized Neil Armstrong's stream of consciousness up to the point when he stepped out of the Lunar Module in 1969.

The lunar example belongs to a large class of phenomenal probabilities that characterize our world. Take the stream of consciousness that you've had over the past hour. Many of this stream's constituent experiences are such that the following is true of each of them: the probability of the stream including that experience where it does, given the rest of the stream's logically independent features, is high, or at any rate greater than chance.[4]

The same is true of many other actual streams. It's also true of many possible streams that never actually occur, like various streams you might have had over the past hour if you'd gone for a walk instead of reading this book.

From the fact that a stream of consciousness with certain phenomenal features is metaphysically possible, nothing follows about the likelihood of a stream having one of those features given that it has its other, logically independent features. The same stream of consciousness can have different associated probabilities in different possible worlds. One of the things that distinguishes the actual world from other possible worlds are the probabilities that are associated in our world with various metaphysically possible streams of consciousness.

Streams of consciousness aren't special in this regard: from the fact that a forest with certain physical features is metaphysically possible, nothing follows about how likely a forest is to have one of those features, such as a certain insect biomass, given that it has its other, logically independent features, such as a certain average tree height.

A probability, phenomenal or otherwise, may or may not have an explanation. But even when something explains why a certain probability exists, the probability is different from what explains it, as shown by the fact that the same probability can have different explanations in different cases. The probability of getting pulled over given that you exceed the speed limit on

[4] Logically independent features are those that don't stand in relations of logical entailment. The fact that a stream of consciousness includes auditory features is logically independent of the fact that it includes visual features, but a stream's including a phenomenally red feature logically depends on its including a phenomenally colorful feature.

a certain road might be low due to the fact that police rarely monitor that road, or it might be equally low because the road is policed by officers who are too lazy to issue speeding tickets. The probability is the same, regardless of what explains it. Likewise, the probability of certain experiences occurring given that certain others occur might be the same in a world in which God decrees that this be the case as in a world in which the probability is due to Kantian noumena having suitable experience-causing powers, or a world in which the probability is just a brute fact.

Most of the phenomenal probabilities that characterize our world are too complicated for us to specify more than vaguely, but that doesn't mean they don't exist with definite values unknown to us. For example, the probability that there's a stream of consciousness culminating in experiences as of a rocky surface, given that there's a stream of consciousness consisting of experiences as of descending deep into Jupiter's atmosphere, is highly uncertain and probably very low, even if Jupiter does have a rocky surface. Experiences as of descending deep into Jupiter's atmosphere are more likely to culminate in experiences as of being killed than anything else, unless maybe the experiences occur in a dream or a flight simulator, in which case who knows what they'd culminate in? Still, there is some conditional probability in this vicinity that has a definite value—something of the form:

> Pr (there is a stream of consciousness culminating in experiences as of p | there is a stream of experiences having features X) = x

where p is a proposition describing the experiences of someone arriving at the Jovian surface, and X the features that such an experiencer's stream might have leading up to his arrival. I can't give you the details, since I've never been to Jupiter, but this doesn't change the fact that there's a conditional probability roughly along these lines having a definite value, any more than my inability to give a detailed description of Neil Armstrong's experiences during the Apollo 11 mission casts doubt on the existence of definite-valued conditional probabilities of the above form related to the experiences he had during that mission.

So that's what phenomenal probabilities are: they're just objective conditional probabilities concerning purely experiential states of affairs. That there are objective phenomenal probabilities that distinguish our world from various other possible worlds is, I think, uncontroversial, and something we're committed to if we think our experience tells us anything about the physical world. In my view, phenomenal probabilities are the building-blocks of physical reality; you don't have to agree with me about that, but you do have to agree that there are phenomenal probabilities, if you think any of our apparently coordinated and non-random experiences are genuinely coordinated and non-random.

I defined the counterfactually possible ideal network as the one our actual experiences most likely belong to in the nearest worlds where our experiences belong to an ideal network, and pointed out that for this definition to work, we need a measure of similarity between possible worlds in something besides physical respects. Phenomenal probabilities provide the required measure: for phenomenalist purposes, the possible worlds closest to ours are those that most closely resemble our world in their phenomenal probabilities.

Above I raised three questions about the possibility postulate. The first, which I've now answered, was: nearest by what measure? (Answer: nearest in terms of phenomenal probabilities.) The second was: why "most likely"? Why not simply define the counterfactually possible ideal network as the one that our experiences belong to in the nearest possible worlds in which they belong to an ideal network?

Because probabilities underdetermine categorical realities. Even if the probability of our experiences belonging to ideal network X conditional on the actual phenomenal probabilities is much higher than the probability of our experiences belonging to ideal network Y conditional on the actual phenomenal probabilities, a world with the same phenomenal probabilities as ours might be one where our experiences belong to Y rather than X. This might be a very unlikely world, but it's still a possible world, and it's among the worlds most similar to ours in terms of their phenomenal probabilities. So if we define the counterfactually possible ideal network as the one that our experiences belong to in *all* the nearest worlds in which our experiences belong to an ideal network, we'll be forced to say that no ideal network is counterfactually possible. The purpose of the "most likely" qualification is to avoid this implication.[5]

The third question I raised about the possibility postulate was: why does it refer to our actual experiences? Why not just define the counterfactually possible ideal network as the one that would (most likely) exist, if there were an ideal network?

[5] Another motivation for the "most likely" qualification is that it lets us avoid Hájek's "might"-conditional objection, discussed above on p. 48, note 59.

Suppose operatives at the National Security Agency have constructed an ultra-powerful supercomputer, which they've programmed to generate conscious experiences just like those that would occur in the ideal counterpart of a possible physical world *W*, different in many physical respects from the actual physical world. To make it vivid, suppose *W* is a world that physically resembles the universe Tolkien describes in his *Lord of the Rings* saga, so that the experiences the NSA computer generates include fully realistic and intersubjectively coordinated experiences as of Ents, dragons, hobbit holes, etc. Call the ideal world that the NSA computer generates *the NSA world*.

The NSA world includes thinglike groups of experiences of Ents, but no thinglike groups of experiences of motorcycles. And, plausibly given our assumptions about the NSA's activities, the ideal world that would most likely exist if there were an ideal world is the NSA world. So if we say that the ideal world for which there's a counterfactual possibility is the one that would most likely exist if there were an ideal world, we'll have to say that the NSA world is the one for which there's a counterfactual possibility. And if we identify the physical world with the phenomenal probabilities in virtue of which a certain ideal world is the counterfactually possible one, we'll have to identify the physical world with phenomenal probabilities grounded in the NSA computer.

We could try to avoid this problem by pointing out that we have no reason to think that the NSA really has a computer like the one described. This is correct, but it doesn't really get to the heart of the problem. We don't want our metaphysics to imply that someone in our world could make it false that there are motorcycles, or true that there are Ents, just by programming a computer the right way. The problem with saying that the counterfactually possible ideal world is the one most likely to exist given the existing phenomenal probabilities is that it saddles phenomenalists with this implication.

We avoid the implication by defining counterfactual possibility partly by reference to our own actual experiences. Even if the NSA does have a computer programmed to generate an ideal world's worth of Tolkienesque experience, that doesn't imply that the NSA world is the ideal world that *our* actual experiences most likely belong to in the nearest possible worlds where they belong to some ideal network. Assuming that our experiences aren't actually part of an NSA simulation, the possible worlds in which our experiences *are* part of an NSA simulation are all very, very far away.[6]

[6] What if we actually are living in a computer simulation? Then the physical world—our physical world, what we mean by "the physical world"—consists of phenomenal probabilities grounded in the computer running the simulation. More on this kind of case in the next chapter, pp. 113–117.

In Chapter 1, I argued that an important advantage of phenomenalism over traditional idealism is that phenomenalism doesn't imply that physical things are mind-dependent. We don't forfeit this advantage by defining counterfactual possibility partly in terms of our own actual experiences. The physical world, in my view, is the totality of phenomenal probabilities in virtue of which a certain ideal world is the one our experiences would most likely belong to, if they belonged to an ideal world. These phenomenal probabilities—heirs to Mill's permanent possibilities of sensation—could exist even if there were no experience, ours or anyone else's. The reference to our own actual experiences in the possibility postulate serves to pick out a certain class of phenomenal probabilities, which I identify with the physical world. But the probabilities thus picked out do not depend on there being any conscious experience.

We can define possible experiences, possible ideal events, etc.—possible in the sense relevant to phenomenalism—as those that occur in the ideal world that our own experiences would most likely belong to, if they belonged to an ideal world. This definition has the advantage of maintaining a continuity with traditional Millian phenomenalism, in which counterfactual conditionals play a prominent role. However, the definition is needlessly complex.

To overcome Chisholm's objection to phenomenalism, we leveraged the modern analysis of counterfactual conditionals in terms of similarities among possible worlds. In order to do so, we took phenomenal probabilities as the principal determinants of similarity: for phenomenalist purposes, similarity among worlds means similarity with respect to phenomenal probabilities.

When we define the possible experiences as those that occur in the ideal world our experiences would most likely belong to if they belonged to an ideal world, we therefore in effect define them as the experiences that occur in the ideal world that our experiences are most likely to belong to, in the possible worlds that are the same as ours in terms of their phenomenal probabilities. But this is just a long-winded way of saying that the possible experiences are those that occur in the ideal world that our own experiences are most likely to belong to, conditional on their belonging to an ideal world.

In view of this, we can stop talking about counterfactually possible experiences, worlds, etc., and instead talk about *probable* experiences, *probable* ideal worlds, etc. Better yet, we can just revert to speaking of "possibilities of sensation" or "possibilities for experience," meaning just the experiences that occur in the metaphysically possible ideal world our own experiences have the highest probability of belonging to, conditional on their belonging to an ideal world. The ideal world for which there's a possibility, in the

Millian sense, is the ideal world that our actual experiences are most likely to belong to, conditional on their belonging to an ideal world; for there to be a possibility for some experience is for that experience to occur in that metaphysically possible ideal world.

5.3 The phenomenalist worldview

When different people observe a rock, their experiences typically relate in certain ways. Phenomenalists see physical reality as a propensity for experiences to occur in such relations, given that they occur at all.

The theory defended in this book is just a detailed working-out of this picture. Conscious experiences tend to occur, when they do, in patterns and relations typical of the experiences that different conscious beings have when perceiving the same physical state of affairs. (Experiences occurring in such patterns and relations are our ultimate evidence for the existence of physical things.) We can describe the experiential tendencies with conditionals of the form: if such-and-such experiential state of affairs existed, such-and-such other experiential state of affairs would exist. The physical truths are conditional truths of this form, and the physical world is the schedule of phenomenal probabilities that makes the conditional truths true.

In Chapter 2, I summed up Mill's phenomenalism with six theses. Here's a similar summary of the phenomenalist theory I favor:

(1) Physical things are possibilities for experience.
(2) As far as we know, nothing grounds, sustains, or otherwise underlies any of these possibilities, except as the case may be other such possibilities.
(3) The possibilities are for ideal objects, events, states, etc.
(4) For there to be a possibility for an ideal thing is for the ideal thing to exist in the ideal world that our own experiences are most likely to belong to.
(5) We perceive a physical thing by having one or more of the experiences that make up the ideal thing for which the physical thing is a possibility.
(6) Some possibilities for experience cause other possibilities for experience, and some cause experiences.

(1), (2), (5), and (6) carry over from Mill's theory essentially unchanged. (3) replaces the quasi-ostensive "experiences that relate to one other as different people's experiences normally do when the people perceive the same thing" with the more precise "ideal events," "ideal objects," etc., understood as defined in the previous chapter. (4), understood against the background of the current chapter's discussion, is basically a version of the corresponding Millian thesis updated to reflect subsequent developments in our understanding of counterfactuals.

The biggest difference between my theory and Mill's is the role that I assign to ideal spacetime. Mill thought he could account for the intersubjectivity of physical things in terms of counterfactual dependencies among possible experiences; I argued that he was mistaken about that, and introduced ideal spacetime to do the work that Mill's theory can't in this regard.

Otherwise, my theory differs from Mill's mainly by including more details about what kinds of experiential structures physical things are possibilities for, and more details about what it is for there to be a possibility for such structures. Even though these details don't mark any major departure from Mill's position, they are important: they disperse the haze that surrounds historical expositions of phenomenalism, and give us the resources to overcome the objections that killed interest in phenomenalism around the middle of the last century. They also provide the background for my next argument for phenomenalism.

6
A Revealing Correspondence

In this chapter, I give a second argument for phenomenalism. Like the first argument, this one parallels an influential argument for materialism, in this case due to U.T. Place and J.J.C. Smart: there is a perfect (exceptionless) correlation or correspondence between conscious experiences and certain types of brain states; the best explanation for this correlation is that conscious experiences just *are* such brain states; therefore, conscious experiences are brain states.[1] The parallel argument for phenomenalism is as follows:

1. There's a perfect correlation between physical things and possibilities for ideal things ("possibilities of sensation," for short).
2. The best explanation for this correlation is that there's no difference between physical things and possibilities of sensation.
3. Therefore, there's no difference between physical things and possibilities of sensation: physical things just *are* possibilities of sensation.

I'll call this the *Correspondence Argument* for phenomenalism, since it hinges on the claim that there's a perfect correspondence or correlation between physical things and possibilities of sensation: a physical thing for each possibility and vice versa.

The first premise of the Correspondence Argument is actually a conjunction of two claims. The first conjunct is that for each possibility of sensation, there is a corresponding physical thing; call this the *sufficiency claim*, since it says that the existence of a possibility of sensation is sufficient for the existence of a corresponding physical thing. The second conjunct is that for each physical thing, there is a corresponding possibility of sensation; call this the *necessity claim*, since it says that the existence of a possibility of sensation corresponding to a physical thing is necessary for the existence of the physical thing.

[1] See Place (1956) and Smart (1959: 142–3); Block and Stalnaker (1999: 23–5).

Phenomenalism: A Metaphysics of Chance and Experience. Michael Pelczar, Oxford University Press.
 DOI: 10.1093/oso/9780192868732.003.0006

To support the sufficiency claim, I build on a broadly Kantian argument that it's sufficient for the existence of any physical thing for there to be something with suitable experience-causing powers: roughly, the power to cause the sort of experiences that people would have if they perceived such a thing; this is the task of §6.1. In support of the necessity claim, I argue, in §6.2, that all physical things are perceivable (albeit not necessarily perceivable by us), from which we may conclude that every physical thing comes with an associated package of possible perceptual experiences of that thing.

The second premise of the Correspondence Argument is that the best explanation for the correlation between physical things and possibilities of sensation is that physical things just *are* possibilities of sensation. Identifying the terms of a perfect correlation is always the best explanation for the correlation, unless there's a strong independent objection to the identification. In §6.3, I consider three objections to the phenomenalist identification, modeled on parallel objections to the materialist identification of conscious states with brain states. I argue that they all fail.

6.1 A thing for every possibility

David Chalmers draws a distinction between skeptical hypotheses and metaphysical hypotheses. A skeptical hypothesis is one "that I cannot rule out and one that would falsify most of my beliefs if it were true." A metaphysical hypothesis is "a hypothesis about the underlying nature of reality."[2]

Consider the *simulation hypothesis* that all our experiences are caused by a powerful computer, as in a classic envatted-brains scenario, or a scenario in which we're parts of a computer simulation like those that Nick Bostrom describes. The computer simulates a physical world in complete detail, and its experience-causing powers are comparable to those of a non-simulated universe.[3]

We can't rule out the simulation hypothesis. Still, according to Chalmers, it isn't a skeptical hypothesis. It's a metaphysical hypothesis that's consistent with my belief that there are tables, chairs, rocks, stars, etc. According to Chalmers, there's nothing in the simulation hypothesis to prevent me from knowing that there are rocks: it's just that if the simulation hypothesis is true,

[2] Chalmers (2010: 457–9).

[3] For Bostrom's discussion, see Bostrom (2003).

rocks have an unexpected underlying nature, in the form of computational states or processes.[4]

Chalmers's stance receives support from the history of science. People's beliefs about the underlying nature of macroscopic physical phenomena have changed dramatically over the centuries, from combinations of the Four Elements, to geometric configurations of Democritean atoms, to dynamical systems of Newtonian bodies, to excitation states of quantum fields. Despite these changes, people's beliefs about the world's macroscopic physical contents have remained highly stable. The ancient Greeks, the natural philosophers of the Enlightenment, and scientists of the 21st century all agree that the world contains rocks, despite having very different beliefs about the underlying nature of rocks.

A plausible explanation for this agreement is that people throughout history have thought that in order for there to be rocks, it's enough that our world has a power to cause the sort of experiences people typically have when they perceive rocks. After all, it's those experiences that everyone throughout history has held in common, as far as their evidence about rocks is concerned. What has changed over the centuries are people's views about what accounts for the experiences.

In Chalmers's view, discovering that we live in a computer simulation would be like discovering that what underlies physical objects are protons, neutrons, and electrons, rather than Earth, Air, Fire, and Water. It would be a discovery about the metaphysical rather than the physical nature of physical things. The physical things are just whatever has the appropriate experience-causing powers, and for a physical thing to have a certain physical nature is just for the experience-causing entity it is to have the power to cause experiences to occur in certain ways: whatever ways experiences typically occur when people perceive a physical thing (thinglike ways, as I've been calling them). Any features the entities with experience-causing power may have over and above their experience-causing powers are irrelevant to the world's physical nature, just as it's irrelevant to a paperweight's functional nature whether it's made of glass or marble.

[4] See Chalmers (2010: 459–73), and for similar arguments, Foster (1982: 191–225), Cargile (1997: 79–80), Huemer (2015: 1050–51), and Smithson (2017). What if we ourselves use a computer to simulate a world containing conscious inhabitants? According to Chalmers, what the people living in that simulation mean by "the physical world" is the computer running the simulation, and what they mean by "the physical facts" are facts about the computer's experience-causing powers. What *we* (the people running the simulation) mean by "the physical world" and "the physical facts" is something else: whatever explains why *our* experiences exhibit regularities suggestive of a physical world.

Let's say that you have maximum justification for believing that p, or, equivalently, that you can be *rationally certain that p,* just in case you're in a position to have (justifiably) the highest credence in p that anyone could rationally have. Being rationally certain that p means being in the best epistemic position that it's possible to be in, with regard to p.[5]

Here is an argument for Chalmers's view, in terms of rational certainty:

1. Rational certainty that there is something with suitable experience-causing powers is sufficient for rational certainty that there's a corresponding physical thing.
2. If rational certainty that there is something with suitable experience-causing powers is sufficient for rational certainty that there's a corresponding physical thing, then the existence of something with suitable experience-causing powers is sufficient for the existence of a corresponding physical thing.
3. So, the existence of something with suitable experience-causing powers is sufficient for the existence of a corresponding physical thing.

Here, by "sufficient," I mean at least *contingently* sufficient: sufficient at least in the sense that being a dog is sufficient for being less than 100 years old, and living on Earth is sufficient for containing carbon. In this sense, *p* can be sufficient for *q* even if *p* doesn't logically or metaphysically entail *q*. If all Fs are in fact Gs, then being an F is sufficient for being a G, in the relevant sense, even if it's not a necessary truth that all Fs are Gs.

As we've seen, the history of scientific discovery suggests that we all already accept the first premise. The premise also receives support from our reflections in Chapter 3 (pp. 52–54, pp. 68–69). It's a premise that most people who aren't skeptics about the external world are likely to accept.

We can support the second premise with a subsidiary inference to the best explanation. Assume, for the sake of argument, that rational certainty that there's something with suitable experience-causing powers is sufficient for rational certainty that there's a corresponding physical thing. The best explanation for this is that the existence of something with suitable experience-causing powers is sufficient for the existence of a corresponding physical thing. (If p doesn't entail q, how can rational certainty that p suffice for rational certainty that q?) So, by inference to the best explanation, and

[5] This isn't the only way to use "certain" as a term of epistemic virtue; for a detailed discussion of epistemic certainty, see Beddor (2020).

discharging the assumption, we can conclude that if rational certainty that something has suitable experience-causing powers is sufficient for rational certainty that there's a corresponding physical thing, then the existence of something with suitable experience-causing powers is sufficient for the existence of a corresponding physical thing.[6]

The conclusion of the argument has an obvious affinity with Kantian metaphysics. Like Kant, Chalmers thinks that our world's physical nature has everything to do with what kind of experiences the world has the power to cause, and nothing to do with why our world has that power. However, also like Kant, Chalmers doesn't offer much detail on what kind of experiences the world must have the power to cause, for its having that power to suffice for the existence of a physical world. Presumably the idea is that the relevant powers are powers to cause experiences to occur the way they do when we perceive some physical entity. But what way is that?

We answered this question in Chapter 4. Kant and Chalmers can say that the relevant powers are powers to cause experiences related so as to constitute an ideal world—a complete network of ideal events located in a common ideal spacetime—and they can identify our world's various physical features with whatever entities have the powers to cause the various ideal events that make up the ideal world for which there is a counterfactual possibility.

If the simulation hypothesis isn't a skeptical hypothesis, what is? Chalmers gives the example of the Chaos Hypothesis:

> I do not receive inputs from anywhere in the world. Instead, I have random, uncaused experiences. Through a huge coincidence, they are exactly the sort of regular, structured experiences with which I am familiar.[7]

Why is this a skeptical hypothesis?

> [W]hat is crucial is that on the Chaos Hypothesis, there is no causal explanation of our experiences at all, and there is no explanation for the regularities in our experience. In all of the previous cases [envatted

[6] Why the appeal to *maximum* justification in the definition of rational certainty? Because even though justified belief that there's a crow in the yard is arguably sufficient for justified belief that there's a black bird in the yard, there being a crow in the yard isn't sufficient for there being a black bird there (albino crows exist). However, having justification for the belief that there's a crow in the yard isn't sufficient for having *maximum* justification for the belief that there's a black bird in the yard, precisely because the presence of a crow in the yard is compatible with the yard's not containing a black bird.

[7] Chalmers (2010: 479).

> brains, computer simulations, etc.], there is some explanation for these regularities, though perhaps not the explanation that we expect. One might suggest that as long as a hypothesis involves some reasonable explanation for the regularities in our experience, then it will not be a global skeptical hypothesis.[8]

The suggestion seems right: as argued in Chapter 3, it's reasonable to think that the physical world is whatever explains why experience tends to exhibit the regularities it does. However, this doesn't settle the question of what explains the regularities. Chalmers's remarks suggest that, like Kant, he thinks of the explanation as causal: the physical world is whatever has the power to cause suitable forms of experience. But the considerations that support the claim that it's sufficient for the physical facts of our world for there to be something with suitable experience-causing powers also support the claim that something less than that is sufficient for the physical facts: namely, that there be the possibilities for experience that Kant and Chalmers posit noumena to ground.

Kant and Chalmers are right that the existence of an underlying substratum with suitable experience-causing powers suffices for the existence of a physical world. But that's because the existence of such a substratum suffices for the existence of suitable possibilities for experience, which are sufficient on their own for the existence of a physical world. In effect, this means that we can keep the powers, but decline to posit any things-in-themselves for the powers to reside in.

We know that experiences occur in seemingly non-random ways, and this gives us some reason to believe that they really do occur in non-random ways. The minimum that has to be true for them to occur non-randomly is that there are suitable experience-related counterfacts or chancy facts—various phenomenal possibilities, as I've been calling them. Maybe the phenomenal possibilities are grounded in noumena with suitable experience-causing powers, but this is more than we can ever know. Since experience is our only source of information about the physical world, and we do know many things about the physical world, it follows that knowledge of experience and possibility for experience suffices for knowledge of physical things.

So we have the following argument for the sufficiency claim:[9]

[8] Chalmers (2010: 479).

[9] Ayer (1936: 53) gives a similar argument, in highly compressed form.

1. Rational certainty that there are possibilities of sensation (possibilities for ideal events, ideal objects, etc.) is sufficient for rational certainty that there are corresponding physical things.
2. If rational certainty that there are possibilities of sensation is sufficient for rational certainty that there are corresponding physical things, then the existence of possibilities of sensation is sufficient for the existence of corresponding physical things.
3. Therefore, the existence of possibilities of sensation is sufficient for the existence of corresponding physical things.

6.2 A possibility for every thing

So far I've argued for the first half of the correlation premise: the half that says that for every possibility of sensation, there's a corresponding physical thing. The second half—that for every physical thing, there's a corresponding possibility of sensation—I argue for as follows: every physical entity is perceivable, and, moreover, perceivable by more than one perceiver; a perceivable entity is, by definition, an entity that's potentially perceived; therefore, for every physical entity, there's a possibility for the sort of experiences that would occur if various perceivers perceived it, related however those perceivers' experiences would relate. It follows that for every physical thing, there's a corresponding possibility of sensation.

The crucial premise of this argument, of course, is that every physical entity is perceivable by more than one perceiver. This is plausible, bearing in mind that "perceivable" here means perceivable in principle (i.e., not *metaphysically* unobservable). As we saw in Chapter 3, phenomenalists have no problem with physical things that no actual being has the capacity to perceive, provided that they are perceivable in principle. Also in Chapter 3, I argued that we're never obliged to acknowledge the existence of metaphysically unobservable physical phenomena. For one thing, interpersonal observability is the main thing that distinguishes actual physical phenomena from fictional physical phenomena, or actual non-physical phenomena (like numbers and dreams). For another, the only evidence that could compel someone to acknowledge the reality of an erstwhile unobserved physical explanatory posit would be evidence that came from actually observing an instance of that posit, or finding out that others have or could have actually observed instances of it, none of which are possible when it comes to metaphysically unobservable things.

In spite of these high-level considerations, you might wonder whether some of the more exotic posits of modern science, particularly quantum physics, are metaphysically unobservable, and, if so, whether this casts doubt on the claim that only metaphysically observable posits have a compelling claim to reality. Let me explain why I think quantum physics poses no threat to the claim that every physical thing comes with a corresponding possibility of sensation.

First, let's bring into focus the quantum phenomenon whose perceivability is in question: quantum superposition. Let *a* and *b* be different points of space, A the property of being located at point *a* (and nowhere else) at a moment *t*, and B the property of being located at point *b* (and nowhere else) at *t*. So, if **p** is a particle,

(1) A**p**

means that at *t*, **p** is located at point *a* and nowhere else, and

(2) B**p**

means that at *t*, **p** is located at point *b* and nowhere else.

One of the main things that distinguishes quantum mechanics from classical mechanics is that it appeals to quantum superpositions, represented by descriptions that look something like this:

(3) $\frac{1}{2}\mathrm{A}\mathbf{p} + \frac{1}{2}\mathrm{B}\mathbf{p}$

One of the main challenges that falls under the heading of "interpreting quantum mechanics" is to say what, if anything, inscriptions like (3) represent. Are there really such things as quantum superpositions? And if so, what sort of thing are they?[10]

This is what John Bell calls the problem of specifying the "beables" of quantum mechanics. Here I'll briefly survey the main proposed solutions to this problem, and argue that they're compatible with the view that all physical things are perceivable.

[10] A more accurate representation of a quantum superposition is: "$\frac{1}{\sqrt{2}}|x, y, z\rangle + \frac{1}{\sqrt{2}}|x,' y, z\rangle$." Also, there are superpositions of features other than spatial location.

The first interpretation, associated with Max Born, Werner Heisenberg, and Niels Bohr isn't strictly speaking an interpretation at all, but a construal of (3) along antirealist lines. According to this interpretation, (3) is simply a recommendation on how to allocate our credences about the location of a particle: (3) tells us to be 0.5 confident that **p** is at point *a*, and 0.5 confident that **p** is at point *b*. And that's all (3) does. By this account, inscriptions like (3) aren't really descriptions or representations at all. They're components of an elaborate symbolic machine whose sole function is to output macroscopic predictions for inputs of macroscopic observations.[11]

Antirealism about quantum superpositions poses no threat to the claim that all physical things are perceivable. If superpositions don't exist, they can't be counterexamples to the claim that all physical things are perceivable.

The second interpretation of quantum mechanics, associated with Louis de Broglie and David Bohm, construes (3) as a partial description of a physical state of affairs in which **p** has a single, definite location (either at *a* or at *b*), but in which the best we can do when forming credences about its location is to be 0.5 confident that **p** is at point *a*, and 0.5 confident that **p** is at point *b*. On the de Broglie–Bohm interpretation, (3) attributes to **p** the property of being a thing for which this is the correct allocation of credences. It's as much a description of our state of ignorance as it is a description of the particle.[12]

The de Broglie–Bohm interpretation doesn't conflict with the claim that all physical things are perceivable. According to de Broglie and Bohm, the truth-makers of quantum mechanical descriptions of the world are just classical physical entities governed by some new laws (represented by quantum mechanical wave-functions). The things that the new laws govern are the same as those that classical physics posits: classical photons, electrons, etc. Since the classical mechanical posits are perceivable in principle (if we have good reason to think they exist at all), the posits of de Broglie–Bohm mechanics are also perceivable, these being the same posits.

According to the third interpretation of quantum mechanics, associated with Hugh Everett and Bryce DeWitt, reality has a branching structure, and (3) describes one fork in that structure. Different versions of this

[11] See Born (1954/1964), Heisenberg (1927/1983: 66, 82–3), and Bohr (1928). Though the antirealist interpretation is widely associated with Born, Heisenberg, and Bohr, it's hard to find any of them endorsing it in a completely unambiguous way.

[12] According to de Broglie and Bohm, to get a complete description of **p** (or its location), we need to supplement the wave-function description of **p** (from which (3) derives) with a pilot-wave description. See de Broglie (1928: 112–19) and Bohm (1952/1983).

interpretation understand the branching structure differently, but the basic idea is that (3) says that from *t* onwards, half of the world's branches are ones in which **p** (or a particle indistinguishable from **p**) is located at point *a* at *t*, and half are branches in which **p** (or a particle indistinguishable from **p**) is located at point *b* at *t*.

In DeWitt's account, different paths through the branching structure are different universes within an all-encompassing multiverse. According to this "many-worlds" interpretation, at *t*, **p** exists both at *a* and at *b*: it's just that scenarios in which **p** exists at *a* occur in different branches of the multiverse from those in which **p** exists at *b*.[13]

The details of Everett's account are debatable, but on one understanding, his idea is that there are many equally complete and accurate ways to describe the physical world, each in terms of a different way of dividing the total state of the world into relevant substates. According to this "relative states" interpretation, we can accurately describe **p** as located at *a* at *t*, and we can accurately describe **p** as located at *b* at *t*, depending on how we describe the total world-system as divided into subsystems. There is, in a sense, no fact of the matter about whether A**p** or B**p**: we can say either (but not both), depending on which of the infinitely many admissible partitions of the world into substates we use when describing **p**. For **p** to be in the superposition that (3) represents is for half of the admissible descriptions to describe **p** as occurring at *a* at *t*, and half to describe **p** as occurring at *b* at *t*.[14]

Both the many-worlds interpretation and the relative states interpretation are compatible with the claim that all physical things are perceivable.

If the many-worlds interpretation is correct, then there are many more physical things than one might have thought, but increasing the number of physical things doesn't make the things any less perceivable.

If the relative states interpretation is correct, then descriptions of **p** as occurring at *a* (at time *t*) relate to descriptions of **p** as occurring at *b* (at *t*) in a way analogous to that in which descriptions of an inertial object (one not subject to any net force) as standing still relate to descriptions of the object as moving in a straight line at constant speed. Just as there's no fact of the matter about an inertial object's state of motion (stationary vs. straight and speedwise constant), on Everett's theory there's no fact of the matter about a particle's location at a given time. This is different from full-blown antirealism about quantum superpositions, since Everett holds that there

[13] DeWitt (1970). [14] Everett (1957).

is a fact of the matter about how the physical world is, in virtue of which it's possible to accurately describe a particle as occurring at *a* at *t* and also possible to accurately describe it as occurring at *b* at *t* (though not possible to describe the particle as occurring at both *a* and *b* at *t*). But there's no reason to think that this underlying physical state of affairs is imperceptible. We might perceive it by having a visual experience as of **p** being located at *a* at time *t*, just as we might perceive an inertial object by having an experience as of an object at rest.

According to a final interpretation of quantum mechanics, discussed by Erwin Schrödinger but not to my knowledge explicitly endorsed by anyone, (3) describes a physical state of affairs in which **p** has its existence equally divided (at *t*) between *a* and *b*: the particle doesn't entirely exist at *a*, and it doesn't entirely exist at *b*, but neither does it entirely fail to exist at either of these points. In a non-toy example, a quantum description of **p**'s location would refer to every point of space in the universe, assigning to each point a value in such a way that the values sum to 1, where the value assigned to a point corresponds to the likelihood of finding the particle at that point if you look for it there. A proponent of the account that Schrödinger alludes to would describe this as a situation in which the particle's existence is unevenly "smeared" over space.[15]

On this quantum smearing interpretation, (3) describes the following state of affairs: it's 50% true that A**p**, and it's 50% true that B**p**; that is, **p** semi-exists at *a*, and **p** semi-exists at *b*. The idea here isn't that half of **p** exists at *a* and half at *b*. It's that all of **p** half-exists at *a* and all of **p** half-exists at *b*.

This is probably the most conceptually challenging interpretation of quantum mechanics. But to whatever extent it makes sense, it's compatible with the claim that all physical things are perceivable. If existence is a gradable property, then it's possible, in principle, to perceive semi-existent things by having experiences with suitably graded qualia. For example, different degrees of existence might show up in a suitably-constituted perceiver's experience as different degrees of phenomenal brightness, ranging from zero (corresponding to complete non-existence) to some maximum phenomenal brightness (corresponding to full existence), with many (perhaps infinitely many) intermediate degrees of phenomenal brightness corresponding to different degrees of physical existence.

[15] See Schrödinger (1935/1983: 156–7).

It seems, then, that the idea of an in-principle unobservable physical entity, implausible to begin with, gets no support from quantum physics.

6.3 An explanation by identification

I've argued that there's a perfect correlation between physical things and possibilities of sensation: that for each physical thing there's a corresponding possibility of sensation, and for each possibility of sensation a corresponding physical thing.

The best explanation for any perfect correlation is always the one that identifies the correlated terms, unless there is some strong independent reason not to accept the identification. You could try to explain the perfect correlation between occurrences of water and occurrences of H_2O other than by saying that water is the same thing as H_2O, but given that there are no objections to identifying water with H_2O, your explanation would be unnecessarily complicated.

What might count as a strong objection to a proposed identification? There are two possibilities. One is an actual counterexample to the identification, like an actual sample of water that isn't H_2O. The other is a good philosophical reason to reject the identification. I've argued that the first objection is not available when it comes to the correlation between physical things and possibilities of sensation. So the question is whether there's any good philosophical argument against identifying physical things with possibilities of sensation.

Since phenomenalism hasn't been discussed much in recent decades, there are no ready-made arguments against the phenomenalist identification of physical things with possibilities of sensation.[16] However, there is a closely related identification that has received a lot of attention: the materialist identification of conscious experiences with brain states. We can look to the ongoing debate over the materialist identity theory for ideas on how one might contest the phenomenalist identity theory.

There are three influential arguments against identifying conscious experiences with physical phenomena (conscious states with brain states, as I'll say for short). One is a modal argument:

[16] Besides Chisholm's argument, which we've already addressed.

1. There are possible worlds that have all our world's physical features, but lack some of the conscious experiences that exist in our world.
2. If the conscious experiences that exist in our world were physical entities (like brain states), such worlds would not be possible.
3. Therefore, the conscious experiences that exist in our world are not physical entities (conscious states aren't brain states).

One example of a world physically identical to ours but without all the conscious experiences that exist in our world is a "zombie world" that contains no conscious experience at all, despite having all of our world's physical features. Another is a world physically identical to ours in which people's visual experience is chromatically inverted relative to our own; in such a world, the experience I have when I see an orange doesn't occur: instead there is a similar but phenomenally blue experience.[17]

The second main argument against identifying conscious states with brain states is the knowledge argument:

1. Someone could know all the physical facts without knowing everything about conscious experience.
2. If conscious states were brain states, it would be impossible for someone to know all the physical facts without knowing everything about conscious experience.
3. Therefore, conscious states aren't brain states.

The classic thought experiment in favor of the first premise is Frank Jackson's Mary, a brilliant scientist who knows all the physical facts, but doesn't know what it's like to have colorful visual experience, because she has lived all her life in a completely black and white environment.[18]

A third, less widely-discussed, but undoubtedly influential argument against materialism is what we might call the argument from what-it's-likeness:[19]

[17] Important versions of the modal argument against materialism are found in Campbell (1970: 100–101), Kirk (1974), and Chalmers (1996).

[18] An early version of the knowledge argument is Broad (1925: 71); the most influential version is Jackson (1982), which is the source of the Mary example. Howard Robinson gives a version of the argument in Robinson (1982: 4–5), and Nicholas Maxwell in Maxwell (1966: 309).

[19] This argument, though rarely explicitly stated, has probably had more influence than any other anti-materialist argument. We find it, or something very similar, in Nagel (1974), and it appears as early as Leibniz (1714/1989: §17).

1. An essential feature of any conscious experience is that there's something it's like to have it.
2. This is not an essential feature of any purely physical phenomenon.
3. Therefore, conscious experiences aren't purely physical phenomena.

All three arguments are contentious, and I'm not going to assess them here. I'm interested in them as models for parallel arguments against the phenomenalist identification of physical things with possibilities of sensation. Let's consider these parallel arguments now.

6.3.1 The modal argument against phenomenalism

First let's consider an anti-phenomenalist counterpart of the zombie argument, which is the most influential modal argument against materialism. I'll call it the Ghost Argument. It goes like this:[20]

1. There are possible worlds with all our world's possibilities of sensation, but no physical entities.
2. If our world's physical features were just possibilities of sensation, such worlds would not be possible.
3. Therefore, our world's physical features aren't just possibilities of sensation.

In support of the first premise, one might argue as follows:

> We can conceive of a world that consists of a multitude of disembodied minds ("ghosts"); call it Ghost World. The ghosts in Ghost World are capable of interaction, and they're disposed to have various experiences when they interact. Due to these dispositions, the ghosts ground the same possibilities for experience that exist in our world. Yet there are no physical objects in Ghost World: it's all just disembodied minds and their experiences. The fact that we can conceive of Ghost World gives us a reason to believe that there are possible worlds that have all our world's possibilities for experience, but in which there are no physical entities.

[20] I discuss this argument in Pelczar (2019: 584–6).

My response is that there is no physical difference between Ghost World and the actual world. (So, I agree with the foregoing argument up to the part that says that there are no physical objects in Ghost World.)

I've already argued that the possibility for a suitable global network of experiences—the possibility of an ideal world—is sufficient for the existence of a physical world. I also argued that the possibility of such a network is sufficient for the existence of a physical world regardless of what, if anything, grounds the network.

In Ghost World, there's a possibility for the same ideal world that there's a possibility for in our world, only grounded in a population of ghosts. Given that it makes no physical difference what, if anything, grounds the relevant possibility, the phenomenal possibilities that the ghosts ground are sufficient for the existence (in Ghost World) of a physical world identical to our own.

Is there something special about Ghost World that prevents the possibilities for experience that exist in it from being sufficient for a physical world just like the one we inhabit? Only if there's a relevant difference between Ghost World and the other possible worlds we've considered in this connection: Berkeleyan worlds, Leibnizian worlds, Kantian worlds, simulated worlds, etc. Unless we can describe a relevant difference between Ghost World and those other worlds, Ghost World isn't a modal counterexample to the phenomenalist identity claim. But there is no relevant difference.

Of course, if our world isn't Ghost World, there must be *some* difference between the two worlds. Phenomenalists recognize this. However, the difference isn't that there are physical things in our world but not Ghost World. It's that in Ghost World, the existence of physical things has an explanation that it doesn't have in ours.

It's notable that no similar reply is available to materialists trying to defend themselves against the zombie argument. Materialists can't plausibly argue that a zombie world does contain conscious experiences, and that the only difference between a zombie world and ours is that in a zombie world, the existence of conscious experience has an unusual explanation. Phenomenalists can plausibly maintain that when we imagine a ghost world, we're just imagining a world in which physical things exist for an unusual reason; materialists can't plausibly maintain that when we imagine a zombie world, we're just imagining a world in which conscious experience exists for an unusual reason.

6.3.2 The knowledge argument against phenomenalism

The second major argument against the materialist identity theory is the knowledge argument. To review: someone, like the fabled Mary, could have complete physical information without having complete information about conscious experience, such as information about what it's like to have phenomenally red experience; if conscious states are just brain states, nobody can have complete physical information without having complete information about conscious experience; therefore, conscious states aren't brain states.

Here's the parallel argument against phenomenalism:

1. Someone (call him Barry) could have complete information about experience without having complete physical information—for instance, without knowing that tomatoes are red.
2. If the physical features of our world are just possibilities of sensation, nobody can have complete information about experience without having complete physical information.
3. So the physical features of our world aren't just possibilities of sensation.

The knowledge argument against materialism assumes that it's possible for someone with complete physical information about our world to lack some information about experience, such as the information, which we do possess, that there's a certain something it's like to have red experience. In order for the knowledge argument against phenomenalism to work, it must be possible for someone with complete information about the experiential facts of our world—facts about which experiences and possibilities for experience there are—to lack some physical information. Otherwise, the first premise of the argument is false.

And it is false. Barry, as described, is impossible. He's impossible, because whatever we know about the physical world can be known on the basis of experiential evidence. So, if Barry really has complete information about experience, he can't know less about the physical world than we do. For example, he can't fail to know that tomatoes are red.

In giving this response, I'm assuming that we can't know anything about the physical world except what's knowable on the basis of experience. Maybe we know some things about the physical world other than on the basis of experience. Some say that we're born with a certain amount of information about the physical world. I take no stand on that. I only maintain that our knowledge of the physical world is limited to what one *could* in theory acquire through experience. (You can think of this as a mild form of empiricism.)

Barry is supposed to lack some physical information that we possess, despite having complete information about experience. But whatever physical information we have is information that one could acquire from experience. Since Barry has complete information about experience—he knows exactly what experiences people have, and exactly what it's like to have those experiences—it follows that he has all the physical information that we have (indeed far more). So he knows, for example, that tomatoes are red.

Barry can't know less about the physical world than we do, since he has all the experiential evidence that our world has to offer, and our own evidence about the physical world is a subset of that. Because of this, the Barry thought-experiment, unlike the Mary thought-experiment, doesn't even get off the ground.

This shouldn't come as a surprise. In Mary's case, we can easily imagine her "Aha!" moment when she first has a colorful visual experience and exclaims: "So *this* is what red experience is like!" What is Barry's "Aha!" moment supposed to be? If he exclaims: "So *this* is what a tomato is like!" what physical feature of the tomato does "*this*" refer to? The tomato's shape? Its color? But Barry already knew all that, since we know it, and we have no more information about tomatoes than he does.

6.3.3 The Johnsonian argument against phenomenalism

James Boswell relates the following encounter with Samuel Johnson:

> After we came out of the church, we stood talking for some time together of Bishop Berkeley's ingenious sophistry to prove the non-existence of matter, and that everything in the universe is merely ideal. I observed, that though we are satisfied his doctrine is not true, it is impossible to refute it. I never shall forget the alacrity with which Johnson answered, striking his foot with mighty force against a large stone, till he rebounded from it, "I refute it *thus*."[21]

[21] Boswell (1799/1953: 333).

Such was Johnson's reaction to Berkeley's idealism. We can only imagine how he'd have reacted to phenomenalism, according to which stones aren't even experiences, but possibilities for experiences: IOUs for ideas of stones, as Johnson might have put it.

The Johnsonian objection is hard to articulate clearly, but it probably exerts more influence than all other objections to phenomenalism combined. Basically, the objection is that mere possibilities of sensation don't have enough "oomph" or substantiality to count as physical objects or events. They're not *thingy* enough to be physical things. But what does this objection really come to?

As far as I can make out, the complaint is that "mere possibilities," whether for experience or anything else, just aren't the right kind of entity to interact with one another, or to be seen, heard, touched, tasted, smelled, or otherwise perceived. They're not the kind of thing you can kick.

The first response to this, paraphrasing Hume, is that anything can cause anything; at least, anything that isn't a mere abstraction. There's nothing in the concept of causation to prevent any non-abstract state of affairs from bringing about any other non-abstract state of affairs. Since the existence of a possibility for conscious experience isn't an abstract state of affairs, we can't prove that stones aren't possibilities for experience by showing that stones cause physical events (like the rebounding of Johnson).

A second response to the oomph-and-substantiality objection is: look at modern physics. Its basic posits—point-events in a non-Euclidean four-dimensional manifold (for general relativity) and complex-valued probability amplitudes (for quantum mechanics)—hardly seem better suited to have physical effects than possibilities for experience. Yet if anything has ontic oomph, the phenomena that physics concerns do.[22]

If possibilities of sensation lack the oomph or substantiality to qualify as physical things, it's evidently not because they're incapable of having physical causes or effects. The other suggestion was that possibilities of sensation aren't the sort of thing we can see, touch, hear, etc. To put the challenge rhetorically: How can you touch a mere possibility?

Phenomenalists have a ready answer: you can touch a possibility of sensation by having tactile experiences that partly realize that possibility.

[22] As Russell wryly comments, "If Dr Johnson had known Eddington's definition of matter [in terms of a tensor that the mathematics of Johnson's day couldn't even represent], he might have been less satisfied with his practical refutation of Berkeley" (Russell 1927: 136).

We touch physical things by having tactile experiences that are among the experiences that those things are possibilities for.

So it looks like the worry that possibilities of sensation are somehow too insubstantial for us to perceive them is also unfounded.

Johnsonians claim that possibilities of sensation don't have the ontic oomph of perceptible and causally potent physical things. But perceptible and causally potent physical things are exactly what the phenomenalist account delivers; or, so I've been arguing. You might as well object that physical things can't be possibilities of sensation, since some physical things are cold and hard. If some physical things are cold and hard, so are some possibilities of sensation, namely the possibilities that are the cold hard things.

This parallels David Lewis's response to a common objection to materialism, which is that it ignores the distinctive feel or what-it's-likeness of conscious experience:

> Far from ignoring questions of how states feel... I have been discussing nothing else! Only if you believe on independent grounds that considerations of causal role and physical realization have no bearing on whether a state is pain should you say that they have no bearing on how that state feels.[23]

In effect: "I've given you states that ground dispositions to engage in avoidance behavior when injured and so forth. What more do you want?"

The problem, for Lewis, is that there's a very natural answer: we want something that we have and zombies don't, and that we know more about than an analgesiac version of Jackson's Mary. Whether this answer withstands scrutiny depends on the outcome of debates over modal and knowledge arguments against materialism, but as long as those debates go unresolved, the answer has bite.

A materialist who claims not to feel something missing from descriptions of conscious experience in purely physical terms may be charged with not taking consciousness seriously; and, as long as we can conceive of unconscious beings satisfying the physical descriptions, or physically omniscient beings suffering from phenomenal ignorance, the charge sticks. A phenomenalist who claims not to feel something missing from descriptions of physical things in purely phenomenal terms might similarly be charged with

[23] Lewis (1980: 222).

not taking *matter* seriously. In this case, however, the charge doesn't stick, since we can't conceive of non-physical things satisfying the descriptions, or phenomenologically omniscient beings suffering from physical ignorance.

I freely admit that the entities to which phenomenalism reduces physical things are on the thin side, ontologically: possibilities for experience, construed here as phenomenal probabilities. But to my mind that's an advantage, as long as they're not so thin as to prevent us from constructing a physical world out of them. That we can construct a physical world out of them is precisely what I've been arguing.

6.3.4 Too much oomph?

The Johnsonian objection is that possibilities of sensation don't have enough ontic oomph to count as physical things. To conclude this chapter, let me address the opposite objection, that in order to make the identification of physical things with possibilities of sensation plausible, I've had to endow the possibilities with *too much* ontic oomph.

I think there are two ways to take this objection.

The objection might be that phenomenalism wrongly, or at least contentiously, treats chance as metaphysically fundamental. Earlier, when discussing the Armstrong doctrine, I argued that we have no good reason to deny that chance is fundamental, and some good reason to think that it is. The point I want to make here is that although phenomenalism is *consistent* with taking chance as fundamental, phenomenalists aren't *committed* to taking chance as fundamental. They're only committed to denying that all facts about chance reduce to *physical* facts. If someone comes up with a successful reductive analysis of chance, we can accept it as a welcome deepening of the phenomenalist position, as long as the analysis doesn't analyze chance in physical terms.

As far as I'm aware, none of the leading contenders for a viable reduction of chance (or "interpretation of probability") analyzes chance in physical terms. This is just as well, assuming, as most people do, that it's at least possible for there to be physically irreducible consciousness. Given that this is possible, we have every reason to believe that chance could exist in a world devoid of physical things. There could be a non-physical world consisting of nothing but low-level sensations—aches, pains, tickles, etc.—that pop into and out of existence, or intensify and abate, with various probabilities, and in which certain types of experiences occur with certain probabilities conditional on the occurrence of other types of experiences. If we were to analyze chance in physical terms, we'd have to deny that such a world is possible.

The other way to take the "too much oomph" objection is that by treating possibilities of sensation as fully real entities with powers to bring about other possibilities (and, sometimes, experiences), I basically just turn my position into an uninteresting relabeling exercise. "OK," the objector says, "fine: you've successfully identified physical things with possibilities of sensation, understood as chancy facts about experience. So what?"

My answer is the rest of this book. In the chapters that follow, I argue that identifying physical things with possibilities of sensation establishes a transparent connection between the world of physics and the world of sense, provides an attractive alternative to structuralist metaphysics, offers a fresh perspective on the problem of consciousness, and yields a satisfying theory of perception, all by taking two things notoriously resistant to reductive analysis—chance and experience—and constructing all else out of them. My hope is that when you see the advantages of identifying physical things with possibilities of sensation, you'll agree that the identification is no mere relabeling exercise, but a substantive and illuminating proposition.[24]

[24] The point doesn't apply just to phenomenalism: any metaphysical theory has to earn its keep by having substantive and illuminating consequences. It's only by having such consequences that structuralism can avoid the charge of merely relabeling physical things "structures," idealism the charge of merely relabeling physical things "ideas," Kantians the charge of merely relabeling physical things "noumena," and materialists the charge of merely relabeling sensations "brain states."

7
Phenomenalism and Science

At first and even second glance, the connection between modern physics and everyday experience is hard to make out. There has to be *some* connection, if the world we perceive is the same as the one that physics describes, but the terms of modern physics are so different from those we use to report our observations in daily life that identifying the connection isn't easy. In §7.1, we consider whether phenomenalism sheds any light on this subject.

A related challenge is to explain how the basic terms of the language modern physicists use to describe the world represent what they do, given that ostensive definitions for such terms are out of the question. In §7.2, I argue that phenomenalism is a plausible consequence of our best account of how basic scientific vocabulary gets its meaning.

Section 7.3 compares phenomenalism to structuralism, an alternative analysis of matter based on a close reading of modern scientific practice. I argue that the main advantage of phenomenalism over structuralism is that it explains more.

7.1 Physics and experience

"Physics and perception," says Russell, "are like two people on opposite sides of a brook which slowly widens as they walk: at first it is easy to jump across, but imperceptibly it grows more difficult, and at last a vast labour is required to get from one side to the other."[1] In this section, I explain how phenomenalism makes the crossing easier.

The basic posits of fundamental physics are, for the most part, imperceptible to us. The challenge is to say how such posits relate to everyday experience, given that the relationship isn't one of perception.

When it comes to humanly unperceivable things, phenomenalists have three options.

[1] Russell (1927: 137).

Phenomenalism: A Metaphysics of Chance and Experience. Michael Pelczar, Oxford University Press.
 DOI: 10.1093/oso/9780192868732.003.0007

First, a phenomenalist can join scientific antirealists in denying that there are any humanly unperceivable things, or (more cautiously) suspending judgement on whether such things exist. Call this "antirealist phenomenalism."

In Chapter 3, we saw that scientific antirealism is plausible as a claim about in-principle unperceivable things, but I don't think it's plausible as a claim about things (like atoms) that are imperceptible to human beings, but not imperceptible in-principle. (Otherwise, the job of defending phenomenalism would be significantly easier.) For this reason, I don't advocate antirealist phenomenalism. (If the balance of considerations ever leans towards scientific antirealism, I'll happily embrace antirealist phenomenalism at that time.)

The second option is to agree that there are atoms, but deny that atoms are humanly unperceivable. Experience is our only basis for believing that any physical phenomenon exists; this is as much true of atoms as it is of sticks and stones. So, the idea goes, why not identify atoms with possibilities for the sort of experiences that justify our belief in their existence: experiences like those that scientists have when performing the experiments or carrying out the observations that lead them to infer the existence of atoms, neutrons, quarks, etc.? Call this "anthropic phenomenalism."[2]

It's hard to find anyone explicitly endorsing the anthropic account, though there are hints of it in the neo-Millian phenomenalists of the early to mid-twentieth century. In any case, the account doesn't stand up to scrutiny.

It would be unreasonable for a blind phenomenalist to think that all physical things are possibilities for the sort of experiences he can have: that would mean either denying the existence of reflections and rainbows, or identifying reflections and rainbows with possibilities for auditory impressions of sighted people's descriptions of reflections and rainbows. Well, I have as much reason to think there could be beings who perceive things I can't as the blind have to think there are people who can perceive things they can't; if I insist on identifying the physical things that such beings could perceive with possibilities for the kinds of experiences I can have, I make the same mistake as the blind phenomenalist.

Furthermore, even though scientists have experiences that (arguably) justify them in positing atoms, there is no individual atom such that any human scientist can have experiences that would justify him in positing

[2] Smart (1963: 27–8) discusses anthropic phenomenalism (though he doesn't call it that).

that particular atom. The humanly attainable evidence is evidence for the existence of atoms as a general phenomenon—atoms in aggregate, so to speak—not for individual atoms. Since atoms can't exist in aggregate unless there are individual atoms, it's hard to see how anthropic phenomenalists can avoid sliding into scientific antirealism.

The phenomenalist's third option is to say that humanly unperceivable things are possibilities for humanly impossible experiences. This is the position I favor. Call it "catholic phenomenalism."

Setting aside scientific antirealism, catholic phenomenalism is, I think, clearly the right position for phenomenalists to take. The phenomenalist's central idea is that physical things are nothing but possibilities for experience—*experience*, not specifically *human* experience. If there are aliens who have experiences very unlike ours, there's no reason why they shouldn't be phenomenalists. If our own perceptual capacities improve through evolution or technological enhancement, that might bring new features of the physical world into view for us, but it won't create new features where there were none before: microscopes didn't make blood cells, they revealed them. Phenomenalists have no more reason than anyone else to tie the world's physical contents to the perceptual limitations of a recently-evolved species of terrestrial ape.

Identifying humanly imperceptible physical things with possibilities for humanly impossible experiences doesn't tell us how the basic posits of physics relate to our own experiences, but it points to a fairly obvious epistemic relationship: we know about the humanly imperceptible posits of physics through their humanly perceptible effects. This is just the standard abductive account of our knowledge of unobservables, supplemented by a phenomenalist analysis of the physical.

But catholic phenomenalism also sees a further, deeper connection between everyday experience and fundamental physical ontology. The things we perceive are possibilities both for the kind of experiences we have when perceiving them and for many experiences that we can't have, and maybe can't even imagine. These different forms of possible experience are related like the levels of a layered hypertext, with possibilities for human experiences linking to possibilities for higher-resolution experiences deeper down in the hypertext. This at any rate is the picture I want to develop here.

Suppose you're looking at a drawing of a 1,000-sided polygon, and an eagle is looking at the same drawing. Since eagles have much better vision than humans, the drawing looks like a polygon to the eagle; to you, it just looks like a circle.

Now suppose the drawing changes (maybe instead of a drawing it's an image on a computer screen). It could be that the changes are such as to result in a change in the eagle's experience (say, if the shape changes from a 1,000-sided polygon to a 999-sided polygon) but not in your experience (to you it still just looks like a circle). However, any change in the geometry of *your* experience would entail a change in the eagle's: for example, if you saw the shape becoming slightly elongated so that it appeared elliptical to you, a corresponding change would occur in the eagle's experience.[3]

When an experience *e* relates to another experience *h* as the eagle's experience relates to yours, I call *e* a *higher-resolution* experience of something that *h* is a *lower-resolution* experience of.

There are many, maybe infinitely many, possible levels of perceptual resolution. Accordingly, a physical thing consists of many, perhaps infinitely many, possible experiences. These different possible experiences are like successive renderings of a zoomable map, like Google Earth, except that you can keep zooming in to individual blades of grass, plant cells, carbon atoms, etc.

It's natural to describe each stage of such a zoom series as amplifying the previous stage. This in turn makes it natural to describe the possible experiences that constitute a physical object as forming a layered hypertext. Each layer amplifies the layer above it, except for the top layer, which consists of possible experiences with the lowest possible resolution.

Identifying physical things with phenomenal hypertexts clarifies the relationship between the world as described by science and the world as revealed to us in ordinary experience: between what Wilfred Sellars calls the Scientific Image and the Manifest Image, or what Russell calls the world of physics and the world of sense, or between Eddington's "two tables": one solid, brown, and static, the "other" a swarm of electric charges zooming around in mostly empty space.[4]

The table I'm sitting at is a hypertext of possible experiences. Some levels of the hypertext are possible experiences as of a solid, heavy, brown, table-shaped object. Other levels are possible experiences as of a mesh of cellulose fibers stuck together with lignin; others still are possible experiences as of a constellation of atoms balanced in mutual equilibrium; and so on. The table is a possibility for the sort of experiences we have when we see a table, for

[3] To avoid Hájek's "might" objection (see above, p. 48, note 59), we can say that if your experience changed, the eagle's experiences would probably change too; or, that the probability of the eagle's experience changing given that yours changes is greater than the probability that yours changes given that the eagle's does. I leave these qualifications implicit in what follows.

[4] See Sellars (1962), Russell (1926; 1927: 130–34), and Eddington (1929: ix–xiii, 247–72).

the sort of experiences that beings who can discriminate individual cellulose fibers have when they see a table, and for the sort of experiences that beings who can discriminate individual atoms have when they see a table. The table isn't *just* the possibility for woodlike experiences, nor just the possibility for meshlike or atomlike experiences: it's a hierarchy of possibilities that includes these possibilities and others.

The view that physical things are phenomenal hypertexts sits well with the commonsense belief that a description of the table I'm sitting at purely in terms of particle physics fails to tell the whole story about the table. It fails to tell the whole story, because it leaves out something that's necessary for *this* table, namely possibilities for experiences of the sort I have when I perceive the table—possibilities that belong to the same phenomenal hypertext as various possibilities for atoms-arranged-tablewise experiences. The possibility for thinglike experiences as of atoms arranged tablewise may be sufficient for the existence of a table (more on this momentarily), but from the fact that there are possibilities for such experiences, we can't infer a priori that there are possibilities for the sort of experiences I have when perceiving tables. But the latter possibilities are as much part of this particular table as possibilities for experiences of atoms arranged tablewise.

We can imagine a world just like ours, except that it has no possibilities for experiences as of atoms; call it Surface World. Surface World contains experiences like those we have when we see, touch, and otherwise perceive tables, and it contains possibilities for many more experiences of that sort that never actually occur, but it contains no possibilities for experiences as of atoms arranged tablewise. We can also a imagine a Skeletal World, just like ours, except that it has no possibilities for the kind of experiences we have when we perceive ordinary physical things. Skeletal World contains various experiences as of atoms balanced in mutual equilibrium, and unrealized possibilities for many more experiences of that sort, but it contains no experiences like those we have when we perceive tables, spoons, trees, etc.

According to the phenomenalist theory I'm defending, neither of these worlds contains the tables that exist in our world. Surface World contains tables, but they're not the tables that exist in our world, since the tables that exist in our world are made of atoms, and those that exist in Surface World aren't (there are no atoms in Surface World). Skeletal World contains atoms. Does it contain tables?[5]

[5] Leibniz (2007: 377–9) raises this question in connection with the Roman Catholic doctrine of transubstantiation. It's not clear where he comes down on it.

This is a question over which I think reasonable people can disagree. It depends on the exact boundaries of one's ordinary physical concepts. My own inclination is to say that there are no tables in Skeletal World. I find that what I mean by "table" is (among other things) something that's capable of manifesting itself in the form of the kind of experiences I have when I perceive a table. I don't think there actually have to be such experiences in order for tables to exist; I don't think there have to be *any* experiences in order for tables to exist. But I think that in order for tables to exist, there has to be at least a *possibility* for the kinds of experiences I have when perceiving tables. So, a world without anything capable of manifesting itself in the form of such experiences is, in my view, a tableless world.

I don't say this of *all* physical things. By "atom" or "quark," I don't mean something capable of manifesting itself in the form of experiences I can have, since I can't have experiences of atoms or quarks. But, for the physical things that I do in fact perceive, and other physical things like them, it's my experiences of those things that are as it were the breeding grounds for my everyday concepts of them (like my concept of a table, a spoon, or a tree). And as far as I can tell, my everyday physical concepts bear the mark of their origins: they apply only to things that are associated with (and, in the phenomenalist view, partly constituted by) possibilities for the kinds of experiences that reveal them to me.

That said, I can see how someone might have somewhat different concepts of various physical kinds. Maybe you think a table is just whatever an ideal completed physics would say a table is, at the most fundamental level. And maybe you think that any world that has possibilities for experiences of things arranged tablewise at that level has tables, regardless of whether it includes possibilities for any of the experiences *we* have when perceiving tables. If ideal completed physics says that tables are states in a high-dimensional Hilbert space, then maybe your view is that if Skeletal World contains suitable possibilities for experiences as of states in such a space, it contains tables—just tables incapable of manifesting themselves in the form of the sort of experiences we have when perceiving tables. Like I said, I'm inclined to think that there are no tables of any sort in such a world, but if you think otherwise, I have no vehement objection. Whatever disagreement we may have on this point arises, I think, from marginal differences in our macro-physical concepts, rather than a substantive disagreement about the underlying metaphysics.

At this point it seems appropriate to say something about the distinction between primary and secondary qualities. I'm not going to say much about it, since I don't think the distinction is useful or important. (It's suspicious that the primary/secondary distinction is extensionally equivalent to the distinction between qualities that we can and qualities that we can't sense in more than one sensory modality; one suspects that if we could feel as well as see things' colors, or see as well as feel their temperatures, we'd count color and temperature as primary qualities.) One thing I will say is that phenomenalism is opposed to the claim that things don't really have secondary qualities: that cherry tomatoes aren't really red, or that fire isn't really hot, but that the redness of a tomato or heat of a flame is really just a quality of the experiences we have when seeing tomatoes or touching flames.

The phenomenalist view is that the redness of tomatoes is a real feature of tomatoes distinct from the phenomenal redness of our visual experiences of tomatoes. For tomatoes to be red is for various experiences, including those we have when we see ripe tomatoes, to be among the experiences for which tomatoes are possibilities. To put it crudely, physical redness is a possibility for phenomenal redness. This doesn't make physical redness unreal, or not a property of tomatoes, or a property of our experiences of tomatoes. By phenomenalist lights, the physical redness of a tomato is an objective, mind-independent feature of the tomato: a feature a tomato has regardless of whether anyone sees it, or whether there is anyone capable of seeing it.

How, in the phenomenalist view, does the redness of a tomato relate to the tomato's surface microstructure? The surface microstructure is a possibility for various high-resolution experiences, like those that beings with much better vision than ours would have if they saw the tomato. These experiences occupy one level of the phenomenal hypertext that the tomato is a possibility for. One level of the hypertext contains experiences characterized by phenomenal redness (as in ordinary human experiences of tomatoes), a deeper level contains experiences characterized by phenomenal micro-iridescence, a still deeper level contains experiences characterized by phenomenal nano-reflectance, etc. The tomato's redness is a possibility for experiences having these various qualia, with the possibilities for phenomenally red experiences occurring higher up in the relevant hypertext than possibilities for experiences as of micro-iridescence, and possibilities for experiences as of micro-iridescence occurring higher up than possibilities for experiences as of nano-reflectance. A being with nano-level powers of perceptual discrimination perceives that tomatoes are red by having experiences that aren't phenomenally red (like ours), but do belong to the same phenomenal hypertexts as the possibilities that our experiences of tomatoes belong to.[6]

While discussing different levels of phenomenal resolution, I've tacitly assumed that physical space and its contents aren't infinitely divided, so that perceiving a physical thing in complete detail requires only finite-resolution experience. What if this assumption is wrong?

Let *H* be an infinitely deep phenomenal hypertext consisting of an infinite series of possible thinglike groups of experiences $\langle g_1, g_2, g_3, \ldots \rangle$ such that for all n, g_n has lower spatial resolution than g_{n+1}. And suppose that one of the groups in this series consists of the sort of experiences that ordinary human beings have when they see a tomato.

There are two cases to consider. In the first case, the series tends towards some geometric limit: there's some geometric form (some shape and size) F such that the shape and size of g_n becomes arbitrarily close to F as n approaches infinity. In the second case, this limit does not exist.

Intuitively, the tomato has a determinate size and shape in the first case, so that as beings with better and better vision look at it, their visual experiences tend to converge on a definite geometric form that the experiences in the series come arbitrarily close to the farther out in the series we go. (The form might be highly scattered, like unconnected points in empty space.) In the second case, where there is no limit, the tomato has no determinate form, so that as beings with ever better vision look at it, their experiences don't tend to converge on any size and shape: maybe the tomato has a fractal-like boundary, and the experiences in the corresponding hypertextual sequence never converge on a single geometric configuration, but eventually start cycling endlessly through a certain subsequence of configurations.

In the first case, where there is a limit, *H* is a well-founded phenomenal hypertext. In the second case, where there is no limit, *H* is a non-well-founded phenomenal hypertext.

Only possibilities for well-founded phenomenal hypertexts are physical things without qualification. If a phenomenal hypertext isn't well-founded, a possibility for it isn't a physical thing simpliciter, but only a

[6] Here I'm speaking of the redness of actual tomatoes. The redness of a tomato in Surface World is just a possibility for phenomenally red experiences like ours: in Surface World, experiences like ours occupy the deepest levels of all phenomenal hypertexts (or possible hypertexts).

physical-thing-relative-to-a-given-level-of-resolution (a given mesh size, so to speak). If some possible phenomenal hypertexts are non-well-founded, there's no such thing as "the physical world," but only the physical world relative to a given level of description. We have no reason to think that our world is such a world, but if it is, this kind of ontological relativization to different levels of description seems to be the right way to understand it.[7]

I've been talking about experiences with different resolutions, but experiences can also differ in other, more radical ways. It's conceivable that there are beings who have non-spatiotemporal forms of experience that we can't remotely imagine, but whose experiences stand in relations of interdependence with our experiences. Such experiences, if there are any, don't fit neatly into the phenomenalist picture I've presented, since, being non-spatiotemporal, they can't occupy the same region of ideal space as other experiences, and so can't be parts of ideal objects or events, as I've defined them.[8]

Regarding the possibility of such alien experiences, phenomenalists have three options.

The first is to say that aliens who have such experiences perceive the same world as us, just not *physical* features of the world. Instead, they perceive non-physical (and non-mental) features of the world that exist alongside the world's physical features.[9]

The second option is to say that the aliens perceive the same physical features of the world we do, despite their experiences not occurring in the

[7] Just as we can distinguish among different spatial resolutions of experience, we can distinguish among different temporal resolutions of experience. If a being with much greater powers of temporal discrimination than mine—a temporal eagle, so to speak—glances at a light bulb that's flashing at the rate of 1,000 flashes per second, it might have an experience of the light flashing 1,000 times, while I, glancing at the bulb at the same time, have an experience of the light shining steadily (no flashes). Had the light been flashing at the rate of 900 times per second instead of 1,000, my experience would still have been of a steadily shining light, but the temporal eagle's would have been of a light flashing 900 times instead of 1,000. The asymmetric dependence between my temporal experiences and the temporal eagle's means that the eagle's experience has a higher temporal resolution than mine.

[8] By "non-spatiotemporal experiences," I mean experiences that lack the phenomenal qualities that make it apt to describe our (human) experiences using spatial and temporal language: qualities like phenomenal length and phenomenal duration. The challenges that possible non-spatiotemporal experiences pose for phenomenalism also arise vis-à-vis the possibility of experiences that are spatial and temporal in the same sense ours are, but have different spatial qualities from ours, such as might characterize the mental lives of aliens whose experiences are systematically spatially deformed (e.g., stretched or skewed) relative to ours.

[9] This is reminiscent of Spinoza's view that the universe consists of infinitely many different and mutually irreducible types of entity, each type being a different "mode" of the single unifying entity, God: see Spinoza (1677/1985: I.16).

same ideal spacetime as ours. The aliens don't have experiences with features like phenomenal length, phenomenal orientation, phenomenal duration, and phenomenal motion; but just as our experiences partially realize a global network of experiences with a worldlike structure, the aliens' experiences might partially realize a global network of experiences with the same general structure. Provided that there are suitable interdependencies among the alien experiences (or possible experiences) and our own, this is arguably a basis for saying that we all perceive the same physical world.[10]

For example, there might be aliens whose experiences present things to them as existing in a high-dimensional Hilbert space like the one quantum physicists use to describe the physical world. Where our experiences have phenomenal features that make it useful to describe them as conscious appearances of states of affairs existing in a space with three geometrical dimensions and one temporal dimension, the aliens' experiences have phenomenal features that make it useful to describe them as appearances of states of affairs existing in a space with a huge number of non-geometrical, non-temporal dimensions. Where our experiences have qualities like phenomenal length and phenomenal duration, the aliens' experiences have phenomenal qualities like nothing we can imagine. But if there's a way to map structural features of the aliens' experiences (or possible experiences) onto the structural features of the experiences we have when perceiving physical things, and if there's an appropriate interdependence between the experiences (human and alien) that exhibit these features, the situation is analogous to one where I perceive the same physical objects as you through visual experiences that are color-inverted relative to yours. In this case, a phenomenalist can say that a physical thing, like a table, includes not only possibilities for experiences as of something existing in an approximately Euclidean space of three dimensions, but also possibilities for experiences as of something existing in a Hilbert space of (at least) 10^{86} dimensions.[11]

[10] A worldlike structure is an abstract structure that a physical world has in virtue of consisting of a certain (possibly infinite) number of basic particulars instantiating certain basic properties and standing in certain basic relations. (For further details, see below, pp. 146–7.)

[11] Which of these possibilities, if either, occupies a deeper layer of the phenomenal hypertext that is the table depends on whether the low-dimensional possibilities asymmetrically depend on the high-dimensional ones (as Albert (2013) suggests), or vice versa (as Monton (2013) suggests). If there's no asymmetric dependence either way, it seems best to say that there's no such thing as *the* space in which physical things exist: they exist in a three-dimensional approximately Euclidean space, by partly consisting of possibilities for experiences (like ours) as of things in such a space, and they exist in a high-dimensional Hilbert space, by partly consisting of possibilities for experiences (like those of the aliens) as of things in a space like that.

If neither of the two options described above is viable, phenomenalists have the third option of simply denying that the aliens experience the same world as us. If there's no interdependence between the aliens' non-spatiotemporal experiences and our experiences, this is certainly the right thing to say. But even if there is interdependence, if it's impossible to relate the aliens' experiences to a worldlike structure to which we can also relate our own experiences, and impossible to construe the aliens as experiencing a non-physical aspect of a world whose physical aspects we experience, there seems no basis at all for thinking that the aliens perceive the same world as us.

7.2 Phenomenalism and scientific language

Suppose, plausibly, that the most complete and accurate description of the world that one could give in the vocabulary of physics would capture all of the world's physical features. (At least, this is plausible if we're talking about an ideal physics that corrects all the errors and omissions of present-day physics.) And suppose it's possible to define the expressions that make up the vocabulary of ideal physics purely in terms of some class of phenomena, X.

On these assumptions, the world's physical features are exhausted by its X-features. (Analogously, we can infer that the world's biological features are exhausted by its microphysical features from the facts (1) that the most complete and accurate description of the world that one could give in the vocabulary of ideal biology would capture all of the world's biological features, and (2) that the expressions that make up the vocabulary of biology are definable in purely microphysical terms.)

What might the X-features be? As a start to answering this question, let's consider how physicists actually define their basic terms. Here's what Eddington says about this:

> The vocabulary of the physicist comprises a number of words such as length, angle, velocity, force, potential, current, etc., which we call "physical quantities." It is now recognised as essential that these should be *defined* according to the way in which we actually recognise them when confronted with them, and not according to the metaphysical significance which we may have anticipated for them. In the old textbooks mass was defined as "quantity of matter"; but when it came to an actual determination

> of mass, an experimental method was prescribed which had no bearing on this definition... You may if you like think of mass as something of inscrutable nature to which the pointer reading [i.e., the prescribed experimental method] has a kind of relevance. But in physics at least there is nothing much to be gained by this mystification, because it is the pointer reading itself which is handled in exact science; and if you embed it in something of a more transcendental nature, you have only the extra trouble of digging it out again.[12]

The idea that all basic terms of physics have, or require, operational definitions came in for a lot of criticism in the decades following Eddington's remarks. But there is a large grain of truth in what he says. Many scientific terms, and apparently all of the most basic ones (apart from abstract mathematical terms) do get defined operationally, and depend for the uses to which scientists put them on being so defined. Not all terms require operational definitions in order for us to use them meaningfully, but operational definitions do seem necessary for the basic terms of physics, insofar as we're in a position to evaluate what physicists use those terms to say.[13]

Since physics employs only a finite number of words and symbols, we can't define them all in terms of physical operations (operations that have physical states as inputs and outputs). You can define "inertial mass" in terms of a function that takes forces as inputs and gives accelerations as outputs, but that leaves "force" and "acceleration" to be defined operationally. You can define "length" in terms of a function that takes certain states of clocks and photons as inputs and gives certain other states of clocks and photons as outputs, but that leaves "clock" and "photon" to be defined operationally. In order for the chain of operational definitions to end without going in a circle, it must terminate in operational definitions that appeal to functions that don't have physical states as inputs and outputs. What kind of operational definition might that be?

Apparently, it must be a definition in terms of *experience:* a definition in terms of functions that take conscious experiences as input and return conscious experiences as output. (Input: experiences as of performing such-and-such an experiment; output: experiences as of getting such-and-such a result.) Call these *phenomenal operations.*

[12] Eddington (1929: 254–5).

[13] For an insightful recent discussion of the role of operational definitions in scientific practice, see Chang (2017).

If this is correct, it follows that the most complete description of the world that one could give in terms of physics is equivalent to some description in terms of possible experiences: the actual or hypothetical inputs and outputs of various phenomenal operations. And if the most complete description of the world that ideal physics can offer captures all of the world's physical features, it follows that the world's physical features are exhausted by the features it has in virtue of containing whatever possibilities for experience it does. But the simplest view in which the world's possibilities for experience exhaust its physical features is phenomenalism. This gives us a good reason to accept phenomenalism.

There are ways to push back against this argument.

You might suggest that operational definitions merely fix the reference of basic terms of physics, rather than giving the terms their meanings, which you might say the terms get by standing in appropriate causal relations to their referents. Maybe mass really is an inscrutable feature of the world, a something-we-know-not-what that operational definitions of mass point to without revealing its essence. The problem with this suggestion, as Eddington notes, is that it's hard to see what's to be gained by positing such an inscrutable something. The posit doesn't seem to serve any purpose, besides that of avoiding the conclusion that physical facts reduce to broadly experiential facts. But unless you're already convinced that this is a conclusion we should avoid, you shouldn't consider avoiding it an adequate reason to posit an inscrutable something.

You might suggest that phenomenalism isn't the *best* view in which basic scientific vocabulary gets defined in experiential terms. At least, there's no obvious reason why a panpsychist or traditional idealist couldn't also define basic physical vocabulary in experiential terms, and you might prefer panpsychism or traditional idealism to phenomenalism. The problem with this suggestion is that it's hard to see how panpsychism or traditional idealism could have any advantage great enough to outweigh the serious disadvantage of entailing that none of our world's physical features would have existed, had there been no conscious experience.

Finally, and more promisingly, you might suggest that basic scientific terms refer to structural features of the world that we can describe without any reference to actual or possible experience. For example, you might suggest that "mass" refers to a property that something has by satisfying some condition statable in purely logical and mathematical terms, or at most logical and mathematical terms supplemented by the verb "to cause."

This is the position known as structuralism (or ontic structural realism). Let's take a closer look at structuralism now.[14]

7.3 Phenomenalism and structuralism

A shared background assumption of everyone who engages in serious theorizing about the physical world's metaphysical nature is that there is a physical world that is as physics describes (subject to future scientific advances). We can say various things about this world, without prejudging any contentious metaphysical questions about it. For example, we can say that it consists of physical particulars that have physical properties and stand to one another in physical relations. Going farther, we can say that it consists of a specific number of particulars that have a specific number of properties and stand in a specific number of relations, and, going farther still, that the properties and relations stand in a particular network of entailments (those that get preserved under any mapping of one set of properties and relations to another).[15] Finally, we can say that the physical world is characterized by specific probabilities; in particular, we can say that certain physical states of affairs have certain probabilities given certain other physical states of affairs.

Now we can focus on an abstract system consisting of a certain number of unspecified particulars having a certain number of unspecified properties and standing in a certain number of unspecified relations, where the properties and relations stand in a certain network of entailments, and certain conditional probabilities hold among various states of the system. We can talk about this system without assuming anything about the further nature of its constituent particulars, properties, and relations. We don't have to think of the system as physical, and we don't have to think of it as experiential. Call this abstract system "the world-structure."

[14] An alternative route to structuralism is via a Ramsey–Lewis definition of physical terms that employs only logical and mathematical vocabulary, or logical and mathematical vocabulary supplemented by the verb "to cause."

[15] By a "mapping," I mean specifically an order-isomorphism, i.e. a mapping that preserves all entailment relations among the mapped properties and relations. A mapping of one set of properties and relations, S_1, to another set of properties and relations, S_2, is an order-isomorphism just in case it satisfies the following two conditions: (1) for any properties ϕ and ϕ' in S_1 and any properties ψ and ψ' in S_2, if ϕ maps to ψ and ϕ' maps to ψ', then $\Box(\phi x \rightarrow \phi' x)$ iff $\Box(\psi x \rightarrow \psi' x)$; and, (2) for any relations Λ and Λ' in S_1 and any relations Π and Π' in S_2, if Λ and Λ' respectively map to Π and Π', then $\Box(\Lambda xy \rightarrow \Lambda' xy)$ iff $\Box(\Pi xy \rightarrow \Pi' xy)$; similarly for relations of more than two places. See Davey and Priestley (2002: 23).

The world-structure is a resource available to idealists, realists, phenomenalists, structuralists, and anyone else who wants to make use of it: it's the common property of all metaphysicians, and any of us can refer to it without circularity or question-begging. This, despite the fact that we came up with the idea of the abstract system by reflecting on our pre-metaphysical notions about the physical world.

Mainstream metaphysical realists identify physical reality with an irreducibly physical realization of the world-structure. Idealists and panpsychists identify physical reality with an experiential realization of the world-structure. Kant identifies physical reality with an otherwise unknowable something that has the power to generate an experiential realization of the world-structure. Phenomenalists identify physical reality with the possibility for an experiential realization of the world-structure. Structuralists identify physical reality with the world-structure itself.

Historically, structuralism was a response to the rarification of physical theorizing in the 20th century. As Eddington remarks, "It was found that science could accomplish so much with entities whose nature was left in suspense that it began to be questioned whether there was any advantage in removing the suspense."[16] Structuralists are among those who think there is no advantage.

According to structuralists, the fundamental nature of the physical world is its structural nature. Physicists produce abstract mathematical models, some of which accurately describe physical reality, or part of it. Normally, we think of an accurate mathematical model as telling us the truth, but not the whole truth, about the aspect of the world it models. According to structuralist metaphysicians, however, that's wrong, or at least it isn't true in general. Structuralists hold that the *best* and most comprehensive mathematical model—the one that an ideal completed physics would give us—*would* tell the whole truth about the physical world. It's not just that physical reality has having-the-structure-of-such-and-such-ideal-mathematical-model as one feature among others. According to structuralism, the nature of physical reality is exhausted by its having that structure. It's structure all the way down.[17]

[16] Eddington (1929: 248).

[17] An early 20th-century defense of structuralism is Russell (1927); see also Paul Dirac's comments in Dirac (1938–9). For more recent defenses of structuralism, see Quine (1981: 16–18), Ladyman and Ross (2007: 130–89), Sider (2011), French (2014), and Tegmark (2014). Probably the earliest example of a modern structuralist position is Boscovich (1763/1922). The general idea arguably goes back to Pythagoras.

The most ambitious versions of structuralism limit the world's structural features to purely logical and mathematical features. Let's call this "thin structuralism": thin, because of the slender reduction base it proposes for the physical world. According to thin structuralism, the nature of a physical thing—any physical thing—is fully described by a statement that uses only logical and mathematical terms: a statement that differs in complexity, but not in basic vocabulary, from "$\exists\phi\,\exists\psi\,\exists R\,\forall x\,\forall y\,((\neg\phi x \wedge \psi y) \rightarrow Rx,y)$," or "$\exists x\,\exists y\,\exists z\,(x \in y \wedge y \subset z)$," or "$\forall x((x > 1) \rightarrow (x^2 > x))$."[18]

Understood in purely logical or mathematical terms, properties are just sets, and relations sets of ordered n-tuples. To define a property in thin structural terms is just to specify the set of things that have it; to define an n-place relation in thin structural terms is just to specify the set of n-tuples of things that stand in it. This opens thin structuralism to the following objection, due to Max Newman.[19]

Given any two universes, U_1 and U_2, in which there exist the same number of things, there's a way of pairing off things in U_1 with things in U_2 so that for any totality of thin structural properties (read: sets of individuals) and any totality of thin structural relations (read: sets of n-tuples), a U_1 thing has one of those properties (i.e., belongs to one of those sets of individuals) if and only if its counterpart in U_2 has a structurally identical property, and a pair (or trio, or quartet, etc.) of U_1 things stands in one of those relations (belongs to one of those sets of n-tuples) if and only if the corresponding pair (or trio, etc.) of U_2 things stands in a structurally identical relation.[20]

The upshot is that any thin structure that one possible world realizes is also realized by every possible world that contains the same number of things as that world. Thin structuralism therefore wrongly implies that it's logically impossible for there to be a world that differs from ours physically, but contains the same number of things.

To escape Newman's objection, structuralists have to thicken their basic ontology to include more than just logical and mathematical structure. The most obvious thickener to reach for here is causal structure. Worlds containing the same number of things are indistinguishable with respect to their thin structural features, but it doesn't follow that they're indistinguishable

[18] Russell (1927) defends thin structuralism.

[19] See Newman (1928: 144–8). Newman's target is the position Russell takes in *The Analysis of Matter*: see Russell (1927: 215, 226–8, 286–9).

[20] Let the entities in U_1 be $x_1, x_2, x_3 \ldots x_n$, and those in U_2 be $y_1, y_2, y_3 \ldots y_n$. Arrange the entities in U_1 into sets, n-tuples, and sets of n-tuples however you like. To get a structurally identical arrangement of the entities in U_2, just replace x_i with y_i, for each integer i.

with respect to their causal structure (provided that we don't construe the causal relation as a mere set of ordered pairs). So, Newman's argument is powerless against what I'll call "thick structuralism": the theory that results from supplementing the basic ontology of thin structuralism with some kind of causal structure.[21]

Thick structuralism is safe from Newman's objection, but many people still find it unsatisfying. The problem is that it's hard to see how structure can contribute to ontology without being the structure *of* something—something that isn't just more structure. Intuitively, the world's structural features underdetermine its physical features. Richard Feynman makes this point at the beginning of a lecture about the nature and discovery of physical laws:

> You think I've told you everything already, because in the lectures I've told you all the great principles that are known. But the principles must be principles about *something*; the principle of the conservation of energy relates to the energy of *something*, and quantum mechanical laws are quantum mechanical principles about *something*—and all these principles added together still don't tell us what the content is of the nature that we are talking about.[22]

The most straightforward way to translate the underdetermination intuition into a sharp challenge to structuralism is with a modal argument: a world could have the same structure as ours without containing any physical things; the possibility of such a world conflicts with the identification of our world's physical features with its structural features; therefore, structuralism is false.

The first premise is the crux of the argument. To support it, opponents of structuralism have to describe a world that duplicates ours structurally, but contains no physical things. This is challenging, since the only non-physical things that have any structure, as far as we know or can even imagine, are abstract entities (as in logic and mathematics) and phenomenal entities

[21] Henceforth, by "structuralism" I mean thick structuralism.

[22] Feynman (1967: 149). The rest of Feynman's comments are somewhat disappointing. The *something*, we're told, is atoms. As for what atoms are: for all Feynman tells us, their nature is exhausted by their being what all the great principles are about. Feynman is far from unique in his skepticism about structuralism: see, among many others, Eddington (1929: 247–72), Hartshorne (1946: 413), Foster (1993: 295), Adams (2007: 40), Strawson (2008: 19–51), Briceño and Mumford (2016), and Goff (2017: 136–40).

(conscious experiences and combinations thereof). Since causation doesn't mix well with abstracta, this leaves purely experiential worlds as the only candidates for modal counterexamples to structuralism.

Accordingly, John Foster envisions a world that perfectly duplicates ours in its structure, but consists of nothing but immaterial, disembodied sensations (tickles, itches, feelings of warmth, etc.). Call this Sensation World. The sensations that make up Sensation World map onto the physical things that exist in our world, in the sense that there's a one-to-one correlation between the sensations in Sensation World and the fundamental physical constituents of our world, such that the structural relations among any collection of sensations in Sensation World are the same as those among the corresponding physical entities in our world. Structuralism implies that Sensation World is physically indistinguishable from our world. But according to Foster, this is wrong: unlike our world, Sensation World doesn't contain any physical things. It's just a network of disembodied sensations.[23]

The problem with this example is that it's open to structuralists to reply that Sensation World *does* duplicate our world physically. They can say that Sensation World is just a non-standard realization of our world's structural features, which shouldn't discourage us from thinking it has the same physical features as our world.

This is similar to how functionalists respond to thought-experiments in which the population of China realizes the same functional features as the brain of an ordinary conscious person: if the Chinese population reproduces the functional features of my brain, then, say the functionalists, it also reproduces the phenomenal features of my conscious mental life. The realizer is different, but the phenomenal state of affairs it realizes is the same.[24]

Appeals to non-standard realizations of actual brains' functional features haven't done much to persuade functionalists to abandon their functionalism, and it seems unlikely that appeals to non-standard realizations of the actual world's structural features will persuade structuralists to abandon their structuralism. Maybe there is some other way to argue that the structural facts underdetermine the physical facts; but instead of pursuing this farther, let's take a closer look at what we get with structuralism, assuming for the sake of argument that it's safe from the threat of underdetermination.

[23] See Foster (2008: 89–93).

[24] See Block (1978) for the "China Brain" argument against functionalism. You don't have to be a reductionist about consciousness to reject this argument; see, e.g., Chalmers (1996: 97, 247–75).

If structuralism works, it offers a simpler analysis of matter than phenomenalism, since it analyses matter into purely structural terms, whereas the phenomenalist analysis calls on possible experience as well as structure (whatever structure characterizes the relevant possibilities). But even if structuralism is superior to phenomenalism when we view the theories narrowly as analyses of matter, if we view them more broadly as theories of matter, mind, and the relationship between them, phenomenalism is superior to structuralism. We've already seen that the phenomenalist analysis delivers a monistic ontology and sheds much-needed light on the relationship between the scientific worldview and everyday experience, and in subsequent chapters, we'll see that phenomenalism offers a fresh perspective on the problem of consciousness and yields an attractive account of perception. This is in stark contrast to the structuralist analysis, which sheds no light on mind or its relationship to matter.

That said, the distance between structuralism and phenomenalism is less than first appears. Like structuralists, phenomenalists are unimpressed by demands to provide physical reality with a categorical nature: a nature over and above what it possesses by having various structural and modal features. (Some structuralists misleadingly describe this as a demand for "objects" or "things.") Structuralists and phenomenalists agree that the nature of physical reality is entirely modal and structural. In this, we are united against idealists, panpsychists, and metaphysical realists. Where we disagree is on the specific modalities: the structuralists take them to be possibilities, probabilities, or potentialities for further possibilities, probabilities, or potentialities, whereas phenomenalists take them to be possibilities, probabilities, or potentialities for experience.[25]

If I wanted to pick a fight, I'd say that structuralist metaphysicians focus too much on what physicists say (better: write) and too little on what they *think*. When you look at what physicists write, what you find is basically a bunch of equations. But why do physicists write down those equations? *To make sense of what they see*, or of what others report having seen. The equations are their attempt to find order and structure *in our experience*.

[25] For structuralists on "objects" and "things," see Saunders (2003: 132): "the logical notion of object is inadequate to ontology... The world is a structure, and it is thought of as such in exact physical, interpreted mathematical terms, but how it is to be broken down into parts, to be spoken of predicatively, can be a more rough and ready affair, sufficient only in the sense of FAPP [for all practical purposes]"; Ladyman and Ross (2007: 130): "There are no things. Structure is all there is"; and French (2014: v): "the fundamental ontology of the world is one of structures... objects, as commonly conceived, are at best derivative, at worst eliminable."

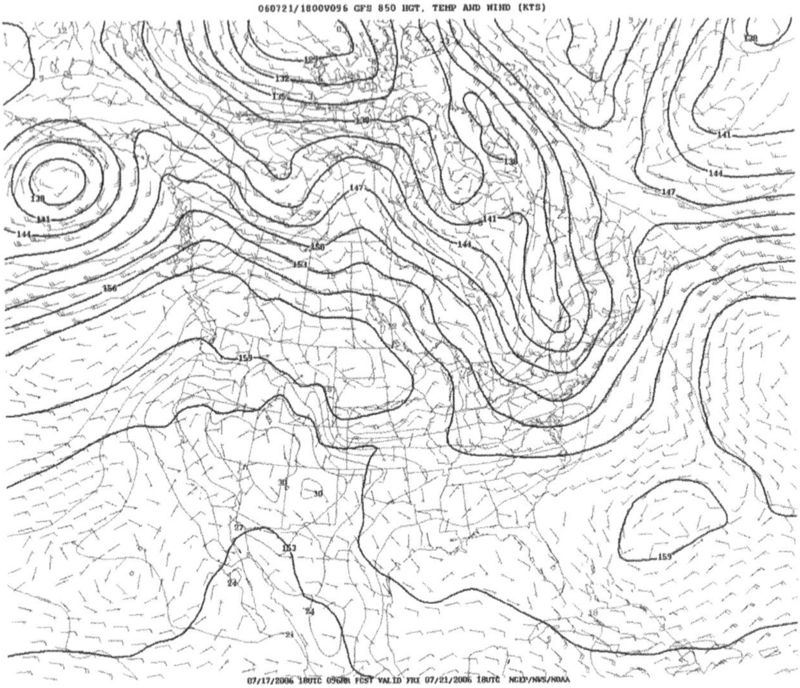

Figure 7.1 Weather map
US National Weather Service

And if they thought there was some better, non-mathematical way to make sense of our experience, they'd forego the math in a heartbeat.

A structuralist metaphysician is like someone who, wanting to give the simplest possible description of the atmosphere, speaks only of variations of pressure and temperature over time and space—all the forces at work in the atmosphere, as represented by the lines and arrows on a weather map—without giving any clue as to what the forces act upon (see Fig. 7.1). If pressed for details, the structuralist will say that whatever the forces act on, it ultimately comes down to more structure: we just have to add more lines and arrows to the map.

But why insist that it's structure all the way down? Whatever we know about the world's physical structure, we know from experience. So experience has to come into the structuralist's picture somehow. Well then, why not take advantage of it by letting experience, or the possibility for it, be what the structure structures? Seen in this light, possibilities for experience give structuralists exactly what they need to complete their picture of the world in a truly satisfying way. They give the structuralist air.[26]

If, like Feynman and Eddington, you think there has to be more to the nature of physical things than that they satisfy the sort of representations found in physics textbooks, it's hard to see where this more is going to come from, if not from the experiential domain. This reflects the fact that our knowledge of the physical world has two ultimate sources: mathematics and experience. Mathematics provides the tools to formulate scientific principles, like the laws of physics, and experience, or the possibility for it, provides something for the principles to be about. Phenomenalism is the most efficient way to incorporate this division of labor between mathematics and experience into a global metaphysical theory: the laws of physics constrain the space of physical scenarios by constraining the space of possible experiences, and the world conforms to the laws of physics by being a possibility for experiences of things obeying those laws.

[26] I've heard David Chalmers describe the structuralist's world as a vacuum containing nothing but winds pushing against other winds: a description that nicely captures both the sublime and the absurd aspects of structuralist metaphysics.

8
Phenomenalism and Consciousness

Phenomenalism is first and foremost an analysis of matter, and its main value is in the light it sheds on the nature of physical things. But phenomenalism also has implications for various topics in the philosophy of mind. We explore these in this chapter and the next.

The topic of the present chapter is the problem of consciousness. Section 8.1 introduces the problem as that of reconciling the prima facie fundamentality of consciousness with its prima facie scarcity. Traditionally, materialists try to solve this problem by saying that consciousness isn't really fundamental, panpsychists and idealists try to solve it by saying that consciousness isn't really scarce, and dualists maintain that the problem is basically unsolvable. Each of these options is open to phenomenalists, and in the pages that follow, we see how each plays out in a phenomenalist context.

8.1 The mystery of consciousness

In a way, consciousness is the least mysterious thing in the world. To those who have it, nothing could be more familiar. Still, there does seem to be something mysterious about consciousness. There are different ways to convey the mystery, or maybe there are different mysteries. The one I want to discuss can be put roughly as follows: given that consciousness is anywhere, why isn't it everywhere?

The mystery arises from two facts, or alleged facts. The first is that consciousness is fundamental: you can't reduce it to, or define it in terms of, anything else. The second is that consciousness is scarce, occurring only in circumstances that are highly localized and, from a cosmic standpoint, extremely uncommon.

The mystery is why something fundamental is so scarce. It's not that there's any contradiction between fundamentality and scarcity. It's just that all other known fundamentals are so ubiquitous that it seems odd that the

Phenomenalism: A Metaphysics of Chance and Experience. Michael Pelczar, Oxford University Press.
 DOI: 10.1093/oso/9780192868732.003.0008

one fundamental thing we know best and care about most—consciousness—would be the unique exception to this rule. In terms of its distribution, consciousness is much more similar to non-fundamental phenomena, like life or brain chemistry, than to phenomena widely accepted as fundamental, like matter and energy.

The problem with consciousness being both fundamental and scarce is that it breaks an otherwise perfect symmetry between fundamentals and foundations: between things we can't reduce to anything else, and things to which we can reduce many other things. You can't reduce a fundamental entity to anything simpler, but foundational fundamentals give us something in exchange for their resistance to reduction: they allow us to reduce other things to facts about them. An entity that's fundamental without being foundational is a dead loss, for explanatory purposes. It resists understanding in terms of anything simpler, without giving us anything in return. It's an ontological couch potato.

It would be different if consciousness didn't exist at all. Then, fundamental or not, it would just be one among many possible but unactualized phenomena, and there would be no mystery surrounding it. Since it does exist, we're left with the question why it isn't pervasive, given that it's fundamental.

There are three ways to approach this mystery.

First, there is the path of *reduction:* deny that there really is a mystery, by acknowledging the scarcity of consciousness, but denying that consciousness is fundamental.

Second, there is the path of *profusion:* deny that there really is a mystery, by acknowledging the fundamentality of consciousness, but denying that consciousness is scarce.

Third, there is the path of *acceptance:* accept that consciousness is both fundamental and scarce, conclude that the mystery is unsolvable, and (optionally) try to make this less disappointing than it sounds.

Materialists take the path of reduction, panpsychists and idealists the path of profusion, and dualists the path of acceptance. Let's see what happens when a phenomenalist ventures down each of these paths.[1]

[1] In this chapter, by "materialism" I mean traditional reductive materialism (which excludes panpsychism), and by "dualism" I mean the traditional dualist view that neither the phenomenal nor the physical reduces to anything more basic (excluding the protopanpsychism discussed in Chalmers 1996: 301–8).

8.2 Orthodox phenomenalism

According to what's probably the most natural version of phenomenalism, and the most traditional, consciousness is both irreducible and scarce, occurring only in association with suitably configured physical systems, like human brains, which are, like all physical things, nothing but possibilities for experience. Call this *orthodox phenomenalism*.

Orthodox phenomenalists follow the path of acceptance. Possibilities of sensation are plentiful: the room is replete with them even when no one's there. Sensations are scarce: there are experiences of the room only when the possibilities for room-experiences are embedded in a larger possibility that also includes possibilities for experiences of a certain type of organism. If experiences reduced to brain states, there would be no mystery here. But assuming they don't, it's puzzling that they occur only in association with possibilities for experiences of certain kinds of brains. What's so special about brains?

Orthodox phenomenalism resembles dualism, in that it considers consciousness both fundamental and scarce. The difference is that orthodox phenomenalism, like all versions of phenomenalism, reduces physical facts to broadly mental facts, namely facts about possibilities for experience. Whether this makes orthodox phenomenalism a kind of monism depends on whether we should classify possibilities of sensation (understood as chancy facts about experience) as mental. On the one hand, they're not as straightforwardly mental as actual occurrent thoughts or experiences; on the other, they have as much claim to mentality as possibilities for physical events (such as chancy facts about particle decay) have to physicality. Since the standard practice is to classify materialist theories of mind as monistic even when they allow for the existence of irreducibly probabilistic facts about the physical, it seems fair to classify orthodox phenomenalism as monistic, even though it allows for the existence of irreducibly probabilistic facts about the mental.

Be that as it may, orthodox phenomenalists definitely agree with traditional dualists in denying that facts about conscious experience reduce to physical facts, and in accepting that despite its fundamentality, conscious experience is scarce in our world. This leaves orthodox phenomenalists in much the same position as dualists vis-à-vis the mystery of consciousness.

Like dualists, orthodox phenomenalists can posit robust correlations between the physical processes that occur in our brains and the conscious processes that occur in our minds, and explain the correlations in terms of natural laws that require certain brain states to occur when and only when certain conscious states occur. But if consciousness is physically irreducible, the existence of such laws is no less surprising or inexplicable than the existence of consciousness itself. This seems true regardless of whether the brain states are irreducibly physical or reduce to possibilities of sensation. Maybe the surprise is slightly less in the phenomenalist scheme of things: after all, if brain states are nothing but possibilities for experience, consciousness comes into the conversation the moment we start talking about brain states, in a way it doesn't in the standard dualist picture. But if orthodox phenomenalism has an advantage over dualism here, it's a minor one. When it comes to the mystery of consciousness, the theories are very much on a par.

8.3 Materialistic phenomenalism

As mentioned in Chapter 1, phenomenalism is compatible with materialism, the view that all mental phenomena, including conscious experiences, are physical phenomena: "conscious states are brain states," in slogan form.[2]

I also mentioned that I'm not inclined to combine phenomenalism with materialism, because of the (to my mind) serious challenges facing materialism. But let's set aside doubts about materialism for a moment, and see how a phenomenalistic version of materialism compares to ordinary non-phenomenalistic materialism.

Materialism has at least one undeniable advantage over all competing theories: it gives a full and relatively conservative solution to the mystery of consciousness (conservative in comparison to the idealist and panpsychist solutions, at least). If consciousness is just a brain process, it's no more mysterious than any other biological process.

Materialistic phenomenalism has the same advantage. It too identifies conscious states with brain states. It differs from standard materialism only by telling a further story about the nature of the brain states: they're possibilities of sensation. Brain states are possibilities for experiences to occur in groups like the one that would exist if a bunch of ideal observers were

[2] Some materialists define materialism as the claim that the mental facts supervene on the physical facts, meaning (roughly) that possible worlds identical in all physical respects are also identical in all mental respects. As far as I can tell, couching materialism in terms of supervenience instead of identity doesn't shield the theory from any of the most serious objections to it (such as those glossed in Ch. 6, pp. 124–125), so I'll continue to speak of materialism as an identity theory.

observing your brain and its physical states right now. Such a group would consist of visual impressions of grey matter, interconnected neurons, various chemical and electrical signals, etc. According to materialistic phenomenalism, the experiences you're having right now just are possibilities for such groups of experiences.[3]

If it seems strange to identify your current experiences with a possibility for the sort of experiences that would occur in people observing your brain, it is—as strange as identifying your experiences with your brain itself (or some of its physical states). This is just the strangeness of materialism. If it puts you off, you should opt for a non-materialist version of phenomenalism.

The point I want to make here is that phenomenalistic materialism has at least one important advantage over standard materialism: unlike standard materialism, which is silent on the nature of the physical, phenomenalistic materialism comes with an analysis of physical phenomena (in terms of possibilities for experience). With standard materialism, you get a solution to the mystery of consciousness. With phenomenalistic materialism, you get that plus an analysis of matter.

8.4 Panpsychist phenomenalism

According to panpsychists, consciousness isn't scarce: it's ubiquitous. In their view, every physical phenomenon, from subatomic particles to the cosmos itself, comes with associated conscious experience. If this is correct, it's no mystery that various states of our brains come with conscious experience.

There are two kinds of panpsychism: weak and strong. Weak panpsychism is the view that every physical entity has phenomenal features. Strong panpsychism is the view that every physical entity is a purely phenomenal entity. According to weak panpsychism, all physical things have phenomenal properties; according to strong panpsychism, all physical things are phenomenal things, and all physical properties are phenomenal properties. Strong panpsychism entails weak panpsychism, but not vice versa.

Weak panpsychism is compatible with the claim that physical things have non-phenomenal properties in addition to their phenomenal properties, and

[3] In order for materialistic phenomenalism to be a formally consistent position, we have to classify probabilistic facts related to experience as topic-neutral. As explained earlier (n. 2, p. 6), whether to classify them as topic-neutral or mental is really just a terminological decision; so, we can formulate materialistic phenomenalism in a way that's at least formally consistent.

so with the claim that our world would contain physical entities, even if there were no conscious experience. According to strong panpsychism, however, physical things have no non-phenomenal properties (aside from abstract structural properties, like the property of having parts, and maybe topic-neutral properties like occurring in time or standing in causal relations). According to strong panpsychism, physical things are literally made of experiences.

Panpsychists' main argument for the ubiquity of consciousness is an inference from our own case.[4] Each of us knows from his own case that some of his physical states (various brain states, it turns out) have phenomenal properties; for example, I know from my own case that some of my brain states have a conscious quality of painfulness, that others have phenomenal color qualities, etc. Nothing about the physical features of these brain states appears to explain why the states have the phenomenal qualities they do, or any phenomenal qualities. The states differ from other physical states (subatomic states, digestive states, meteorological states, etc.) in many physical respects, but none that suggests that the brain states rather than the other states come with conscious experience. Whatever physical features distinguish the brain states from other physical states, and whatever physical roles the brain states play in the systems they occur in, there's no evident reason why states with those physical features and roles should have conscious qualities, and states with different physical features or roles should not. If it weren't for the observed correlations between our own conscious experiences and our own physical states, there'd be nothing to suggest to us that any of our physical features had conscious qualities.

At this point, we can go in either of two directions.

From the fact that we only ever find evidence of consciousness in conjunction with certain kinds of brain states, we might infer that consciousness occurs only in conjunction with such states, despite the states not differing from other physical states in any way that suggests that they're conscious.

Alternatively, from the fact that our brain states don't differ from other physical states in any way that suggests that the brain states are conscious, we might conclude that *not* only the brain states but *all* physical states are conscious, and account for the fact that we only ever find direct evidence of consciousness in conjunction with certain brain states as due to the fact that our own consciousness happens to consist of (or at least inhere in) such

[4] See Eddington (1929: 258–60).

brain states: those brain states happen to be the chunks of conscious reality that our minds are (or, inhere in).

Panpsychists favor the latter hypothesis. If the hypothesis is correct, a proliferationist solution to the mystery of consciousness follows directly. I'm not going to weigh in here on whether the hypothesis is correct. The point I want to make is that phenomenalism is compatible with it, at least in its weak panpsychist form.

According to phenomenalism, all physical things are possibilities for experience, but phenomenalists traditionally hold that relatively few physical things come with any associated experiences. The exceptions are physical things that get perceived by conscious beings, and the physical things that are the neural correlates of conscious beings' experiences. Perceived things come with experiences that (partially) realize the possibilities for experience that the things are; the neural correlates of conscious experiences (various brain states) come with experiences that don't realize the possibilities for experience that the brain states are, but are linked to those states by relations of dependence.

It's the latter sort of experiences that panpsychists say characterize all physical things. For example, when I'm in pain, my brain is in a physical state, S, which is the neural correlate of my pain. This state, according to phenomenalists, is a possibility for (among others) the sort of experiences a neuroscientist would have if he perceived S. My pain is not among the experiences for which S is a possibility, but it is linked to those possible experiences, inasmuch as it's an effect, or at least a non-accidental concomitant, of the brain state (S) that is the possibility for those experiences (various experiences of a brain). Likewise, when I see a tree, my brain is in a certain state, which is, according to phenomenalists, a possibility for (among others) the sort of experiences a neuroscientist would have if he perceived that brain state. My visual image of the tree isn't among the experiences for which the brain state is a possibility, since it has a different location in ideal space from those experiences, but it is linked to the experiences, inasmuch as the image is an effect, or at least a non-accidental concomitant, of the brain state.

As these examples illustrate, phenomenalists recognize that the neural correlates of our conscious states come with phenomenal qualities that don't characterize the experiences for which the neural correlates are possibilities. Endorsing weak panpsychism just means saying that *all* physical entities have such qualities. C-fiber firings come with painful qualia that don't characterize the experiences that C-fiber firings are possibilities for; electrochemical events in the brain's visual cortex come with visual qualia that don't characterize the experiences that the electrochemical events are possibilities for;[5] and, according to the panpsychist phenomenalist, sticks and stones come with qualia that don't characterize the experiences that sticks and stones are possibilities for.

Even though weak panpsychism already offers a proliferationalist solution to the mystery of consciousness, most panpsychists today don't stop at weak panpsychism, but make the strong panpsychist claim that physical things don't merely have conscious experiences, but are purely conscious entities. Why do they take this further step?

To secure physical things intrinsic categorical natures. The reasoning here is: physical things must have categorical natures; the only categorical natures we know of are phenomenal natures; all physical things do have phenomenal natures (per weak panpsychism); so, it's reasonable to conclude that the categorical natures of all physical things are exhausted by their phenomenal natures.

This reasoning invites several objections, but here I want to raise just one, which is that we have no good reason to accept that physical things have categorical natures.

In Chapter 3, we considered and rejected several arguments for the Armstrong doctrine that physical things must have categorical natures. Panpsychist arguments for the Armstrong doctrine rely heavily on metaphors and analogies. Here is a typical example:

> [T]here is a basic intuition that causal powers are too *metaphysically thin* to constitute the complete nature of fundamental concrete objects. A causal power concerns how its bearer points toward other entities and toward its own non-actual but potential manifestation in reaction to those entities. But, intuitively, fundamental objects must also have a *manifest nature:* a nature that does not consist in such shadowy pointing, but consists in how the object is *in and of itself*. This view is commonly expressed with the analogy that a world of pure powers is like a world in which things are continuously packing their bags for a journey that is never taken: objects continuously change their potentialities, but those potentialities never result in anything *actual.*[6]

[5] Unless the electrochemical events happen to be the neural correlates of someone's experiences of those very events, as might happen if a neuro-scientist observes his or her own visual cortex.

[6] Goff (2017: 140). For similar remarks, see Eddington (1929: 262), Hartshorne (1946: 413), Foster (1993: 295), Adams (2007: 40), and Strawson (2008: 20).

As we saw in our earlier discussion of ontic structural realism, it's not clear that a purely structure- or powers-based conception of the physical universe is as indefensible as this quote suggests. But set that aside. The point I want to make here is that you can hold that physical things don't have (or at least don't require) categorical natures, without holding that physical things are nothing but powers to produce further powers. You might be a Kantian who holds that physical things are powers to produce—not further powers, but—*experiences*; or, you might a phenomenalist who holds that physical things are possibilities for—not further possibilities, but—*experiences*.

The intuition that causal powers (or dispositions, or possibilities) are too metaphysically thin to constitute the complete nature of a physical thing is strong, as long as you confine your attention to powers, dispositions, and possibilities unrelated to experience. But the intuition loses much of its strength when we realize that powers, dispositions, and possibilities can include powers or dispositions to cause experiences, or possibilities for experiences. It's hard to see how a physical thing could be nothing but powers or possibilities, if the powers are taken to be nothing but powers to produce further powers, and the possibilities nothing but possibilities for further possibilities. But if the powers are allowed to include powers to cause experiences, or the possibilities to include possibilities for experience, it becomes much easier to see how a physical thing might consist entirely of such powers or possibilities. And we can admit such powers or possibilities into our ontology without identifying physical things with conscious experiences. We can be Kantians or phenomenalists, instead of strong panpsychists.

I've argued that the rationale for advancing from weak to strong panpsychism is questionable. There are also problems with strong panpsychism itself.

For one, as noted earlier, strong panpsychism implies that none of our world's physical features could exist in a scenario devoid of experience. After all, if it's in the Moon's very nature to be an experiential something, the Moon can hardly exist in a world in which there's no experience. Yet, it seems easy to imagine the Moon existing in a world in which there's no experience (e.g., a world in which sentient life never evolves). This isn't a knock-down argument against strong panpsychism, but it raises doubts about it similar to those that the zombie argument raises about ordinary reductive materialism.

Another problem for strong panpsychism is familiar from our earlier criticism of theories, like metaphysical realism, that try to identify physical things with categorical entities that allegedly ground our world's possibilities for experience, rather than with the possibilities themselves. Imagine a world consisting of all the experiences that strong panpsychists identify with physical reality, but in which those experiences don't support any of the sensation conditionals that hold in our world. The experiences have no power or disposition to cause experiences as of physical objects and events occurring in a common time and space, and ground none of the experience-related counterfactual conditionals or conditional probabilities that characterize our world. Intuitively, and as argued in Chapter 6, the described world doesn't contain the same physical things as our world (or even, perhaps, any physical things at all). Yet, strong panpsychists are committed to saying that such a world would be exactly like ours, physically, since it would contain everything that they say our physical world consists of.

It's not surprising that strong panpsychism is open to the same objections as metaphysical realism: strong panpsychism basically just is metaphysical realism, except that it replaces the realist's non-mental substrata with mental substrata.

According to phenomenalists, our world's physical contents are fully determined by its experiential possibilities, and not by what, if anything, underlies them (whether the ostensible underlying entity is a God, a population of monads, or a complex of panpsychist experiences). Phenomenalism isn't compatible with strong panpsychism, any more than with metaphysical realism. This is no great loss to phenomenalists, however, since they're still free to adopt weak panpsychism, which offers the same solution to the mystery of consciousness as strong panpsychism, at a much lower cost.

8.5 Panoptic phenomenalism

Panpsychists hold that all physical things have (weak panpsychism) or are (strong panpsychism) conscious experiences, but they don't say much about the nature of those experiences, besides that it's unlikely that they resemble ours. (Who knows what kind of experiences doormats and electrons might have?)

In contrast, idealism offers a proliferationist solution to the mystery of consciousness that says quite a lot about the nature of the (many) experiences it posits. They're experiences of the sort sentient beings have when they perceive physical things.

According to idealists, every physical entity comes with conscious perceptual experiences of that entity, since every physical entity *is* conscious perceptual experiences of that entity.[7] In the idealist view, our world actually contains all the experiences that it would if there were all the ideal observers described in Chapter 1. For every physical object and event, there are experiences of that object and event as perceived from every distance and angle; for every path through physical spacetime, there are streams of consciousness like those that would occur in conscious observers who followed that path. This ideal world isn't a mere possibility, according to idealists, but an actuality of which our own experiences are just one small part.

Suppose that this ideal counterpart of our physical world does actually exist. Idealists say that the ideal world just *is* our physical world, but we can suppose that the ideal world exists without making that identification. As phenomenalists, we can continue to hold that the physical world is the possibility for the ideal world, not the ideal world itself. In the phenomenalist view, to suppose that the ideal world actually exists is just to suppose that all of the possibilities that constitute the physical world are wholly realized (in the form of actual experiences). This is the view I want to explore now. Call it *panoptic phenomenalism*.

Strange as it may seem at first, there's something appealing about the panoptic picture. Here we have a vast network of conscious experiences comprising innumerable streams of consciousness. Some of the experiences are mine, some are yours, but most occur without our having any inkling of them. It may seem odd to think that there's so much experience in the world, but if physical things are nothing more than possibilities of sensation, there's also something odd about the idea that only a small subset of those possibilities get realized in the form of actual experience. If realization is the default for possibilities of sensation in our world, this oddity disappears.

Like idealism, panoptic phenomenalism yields a proliferationist solution to the mystery of consciousness: if our universe contains an ideal world's worth of experience, experience is the opposite of scarce. However, the solution might seem to trade one mystery for another.

If our world contains all the experiences that the panoptic view posits, the vast majority of them are evidently unembodied, or at least not embodied in any form we can discern. This gives rise to a sort of inverse problem of

[7] Recall Berkeley: "a certain colour, taste, smell, figure and consistence having been observed to go together, are accounted one distinct thing, signified by the name apple; other collections of ideas constitute a stone, a tree, a book, and the like sensible things" (Berkeley 1710/1901: §1; 1710/1982).

consciousness. Now the question isn't, "What's so special about bodies like ours that they alone should come with conscious experience?" but: "What's so special about conscious experiences like ours that they alone should come with bodies?" More generally, how come our perceptual experiences have physical (neural) correlates, when the vast majority of the perceptual experiences that we're supposing to occur—perceptions of the roots of mountains, individual atoms, events occurring in the primordial past, etc.—don't?

Fortunately, the inverse problem of consciousness has a straightforward solution. Given that all the possible experiences and streams that characterize an ideal counterpart of our physical world actually exist, and given that our physical world includes animal bodies moving through time and space and interacting with their physical environments the way they do, it's inevitable that there are perceptual experiences of various physical features of my body, its environment, and their physical relationship, including but not limited to my own experiences of these things. Given that our universe actually contains all the experience that exists in an ideal counterpart of the physical world, the fact that some but not all experiences of trees occur in tandem with experiences of animal bodies interacting with trees is no more surprising than the fact that some but not all trees occur in tandem with animal bodies interacting with trees. In a panoptic scenario, it's no mystery that our experience is embodied when most experience isn't, any more than it's a mystery that some lottery tickets are winners when most aren't.

Suppose that the default condition for human beings is life in solitary confinement. Most people are born and die in solitary, without ever coming into contact with another human being, or even knowing that there are other human beings. (Maybe machines are in control.) Once a week, there's a lottery. Every prisoner gets a ticket, and out of all these billions of tickets, there's just one winner every week. The prize is to be released from prison into a small community of previous lottery winners living in freedom outside. The only way to get out of prison is by winning the freedom lottery.

Well, the only way for a conscious being to have embodied experience like ours is by winning the animal embodiment lottery: i.e., by being one of the small minority of conscious beings whose perceptual experiences include both perceptions of trees, tables, etc. and various perceptions and proprioceptions of an animal body related in certain ways to trees, tables, etc. As a winner, you get to observe various correlations between your experiences and various happenings in and around a certain body (yours). This puts you in a position to infer (defeasibly, but confidently) that beings with bodies like yours have experiences like yours. (Neither the observed correlations nor any other information you have suggests that conscious experience occurs except in beings with bodies similar to yours.)

Winning the animal embodiment lottery is like winning the freedom lottery. If you win the freedom lottery, there's nothing surprising or mysterious about the fact that everyone you know of is a freedom lottery winner: if they hadn't won the lottery, you'd never have known they existed. Likewise, if you win the animal embodiment lottery, there's nothing surprising or mysterious about the fact that every conscious being you know of has won the same lottery: if they didn't have animal bodies, you wouldn't know that those conscious beings existed.[8]

Like idealism, panoptic phenomenalism denies that consciousness is scarce, by positing a prolific amount of perceptual experience: all the experience for which the physical world is a possibility. This is enough to solve the mystery of consciousness as stated above, but it leaves a related question unanswered: why is it that certain physical states do, but the vast majority seemingly do not, come with non-perceptual phenomenology (pains, tickles, pangs of regret, feelings of euphoria, etc.)? Even if perceptual experience is ubiquitous for the reasons idealists and panoptic phenomenalists give, it doesn't follow that non-perceptual experience is ubiquitous. Since non-perceptual experience has as much claim to fundamentality as perceptual experience, it would be mysterious if non-perceptual experience were as scarce as commonly supposed.[9]

To address this, a phenomenalist can combine the idealist's proliferation of perceptual experience with the panpsychist's proliferation of non-perceptual experience. In the view that results, each physical entity comes with conscious perceptions (all of those for which the entity is a possibility), and each physical entity has phenomenal properties analogous to those that characterize the neural correlates of our own experiences, including non-perceptual

[8] Here you might wonder whether there isn't more *embodied* experience than commonly supposed. Maybe rocks are the bodies of conscious beings who have won (?) the mineral embodiment lottery, and likewise for all other physical things. Still, it seems unlikely that there's enough physical stuff in the world to embody an entire ideal world's worth of experience.

[9] Similar remarks apply to non-veridical perceptual experiences, like dreams and hallucinations. More on these in the following chapter.

experiences like pains and feelings of euphoria. We might call the version of phenomenalism that combines these ideas *panexperiential phenomenalism*. It combines the essence of the panpsychists' and idealists' solutions to the mystery of consciousness, while avoiding the panpsychists' and idealists' commitment to a mind-dependent conception of physical reality.

9
A Phenomenalist Theory of Perception

Phenomenalism comes with a simple theory of perception: to perceive a physical thing is to have one or more of the experiences for which the thing is a possibility. In this chapter, I elaborate on this theory and compare it to other theories of perception.

Section 9.1 reviews three questions that shape current philosophical debate about perception. Section 9.2 considers how the two leading theories of perception—representationalism and naive realism—answer those questions. Section 9.3 gives the phenomenalist answers to the questions. Section 9.4 compares the phenomenalist theory of perception to two additional theories: a pre-theoretical folk theory of perception, and a Berkeleyan theory.

A note on terminology: throughout the following discussion, by "see," "hear," "perceive," etc., I mean have an accurate or veridical experience of something. (I use "veridical" and "accurate" interchangeably.) In my terms, you can't perceive what doesn't exist, and inaccurate or non-veridical perceptual experiences aren't perceptions of anything: you don't see or hear anything in dreams, even if you have perceptual phenomenology indistinguishable from what you might have when seeing or hearing something. Only accurate or veridical perceptual experiences are perceptions *of* anything. I'll continue to use the phrase "as of" to indicate the phenomenal character of perceptual experiences, whether veridical or non-veridical. When you dream of walking in the woods, you have experiences as of trees, even though you don't perceive (see, hear, feel, etc.) any trees.

9.1 Three questions about perception

Three questions animate the philosophical debates about perception that are most relevant to phenomenalism. First, there is the Accuracy Question:

How do accurate perceptual experiences differ from inaccurate ones?

Phenomenalism: A Metaphysics of Chance and Experience. Michael Pelczar, Oxford University Press.
 DOI: 10.1093/oso/9780192868732.003.0009

Many perceptual experiences give us accurate information about the world; others, including the experiences we have when dreaming or hallucinating, or when subject to perceptual illusions, do not. What distinguishes the accurate experiences from the inaccurate ones?

Second, there is the Diversity Question:

> *How can phenomenally different experiences be accurate perceptions of the same state of affairs?*

When I see a moth flitting by in the moonlight, I perceive its motion by having visual experiences. Meanwhile, a bat perceives the same motion by having echolocational experiences. In virtue of what are our experiences perceptions of the same thing?

Third, there is the Specificity Question:

> *How does perceiving that* p *differ from knowing that* p *in a non-perceptual way?*

Seeing that there's a moth flitting by is different from believing that there's a moth flitting by on the strength of adequate but non-perceptual evidence, such as testimonial evidence. It's also different from believing there's a moth flitting by on the strength of adequate non-perceptual evidence accompanied by a vivid hallucination of a moth flitting by. But what exactly is the difference?

Before looking at how the two leading theories of perception answer the Accuracy, Diversity, and Specificity questions, I should mention a fourth question that also looms large in current philosophical debates about perception: "Why do perceptual experiences have the particular phenomenal characters they do?" Call this the Character Question.[1]

I'm not going to discuss the Character Question. Since I think consciousness is fundamental, I don't think there can be a reductive explanation for why perceptual experiences have the phenomenal characters they do. If a reductive explanation is possible, a phenomenalist can accept it—as argued earlier, phenomenalism is compatible with reductionism

[1] For the Character Question, see, e.g., Harman (1990), Shoemaker (1994), Tye (1995: 93–159), Byrne (2001), Loar (2003), Pautz (2010), Millar (2014), and Langsam (2020).

about consciousness—but phenomenalism certainly doesn't presuppose a reductive view of consciousness, so the Character Question is largely a side-issue for phenomenalists.[2]

9.2 Representationalism and naive realism

What is it, to see a red tomato? Before tackling this question, let me make a couple of working assumptions. First, I assume there's no difference between seeing a red tomato and seeing that there is a red tomato. Second, I assume we can safely set aside questions about just how much you see when you see a red tomato: the whole tomato, or just the part of it facing you, for example. These issues would deserve close attention in a detailed assessment of theories of perception, but here I'm only trying to get the mainstream options on the table, and for that we can skate over these finer points.

9.2.1 Representationalism

What is it, to see a red tomato? Here's one answer: to see that there's a red tomato is for a red tomato to cause you to have an experience that represents that there is a red tomato.

This is the answer we get from representationalists about perception.[3]

Representationalism is so-called because it construes accurate perceptual experiences as phenomenal signs or representations of their causes. To state the representational theory more precisely would require saying more about the representation relation that accurate perceptions ostensibly bear to their objects, and more about exactly which cause of a perceptual experience representationalists mean by "its cause." I'm not going to go into any of

[2] If there's a non-reductive explanation of perceptual phenomenal character in terms of a lawlike correlation between the phenomenal qualities of perceptual experiences and what the experiences represent, it's also available to phenomenalists, who have no special reason to deny that perceptual experiences represent, or that there are laws connecting experiences' representational contents with their phenomenal characters.

[3] Among representationalists there's a division between "sense datum" theorists and "representational content" theorists. According to representational content theorists, a perceptual experience's representational content explains its phenomenal character: see, e.g., Harman (1990) and Tye (1995: 93–159). Sense datum theorists don't try to explain phenomenal character in terms of representational content: see Chisholm (1957: 142–67), Anscombe (1965), Armstrong (1968: 208–44), Pitcher (1971), Jackson (1977), and Robinson (1994).

that here. Instead, I just want to review the prima facie advantages and disadvantages of representational theories.

Representationalism has a straightforward answer to the Accuracy Question. An accurate perceptual experience is one that's caused by the state of affairs it represents; you have an accurate perception of a red tomato just in case the presence of a red tomato causes you to have an experience that represents the presence of a red tomato. An inaccurate perceptual experience is one that's not caused by the state of affairs it represents. If you hallucinate or dream of a tomato, your experience represents that there is a tomato present, even though there is no tomato present to cause your experience. Similarly, according to representationalists, when you look at Fig. 4.3 on p. 91, your experience represents the squares marked A and B as being different shades of gray, even though there's no color difference causing your experience (the regions of the page causing your experience have the same color).

Representationalism also has a plausible answer to the Diversity Question. When I see a moonlit moth, the moth's presence causes me to have an experience that represents that there's a moth flitting by. When the bat echolocates the same moth at the same time, the moth's presence causes the bat to have an experience that represents that there's a moth flitting by. The difference between the phenomenal quality of my experience and the bat's is like the difference between depictions of the same thing in different media.[4]

The main challenge for representationalists is how to answer the Specificity Question.

Imagine that you're getting your information about the moth from a live radio broadcast. It's David Attenborough commentating, and his commentary is so good that it's almost like you're there, seeing the moth flitting past in the moonlight. If your imagination is good enough and you've seen this kind of moth before, you might have vivid mental images of the moth as you listen to the broadcast. If your imagination is *very* good, or you've taken an imagination-enhancing drug, you might even have visual phenomenology indistinguishable from the phenomenology that Attenborough himself is having.

Suppose that your imagination is that good, or that you have taken the drug. As you listen to the broadcast, you form visual images of the moth

[4] This is the answer we get from sense datum representationalists. Representational content theorists have a harder time explaining phenomenal differences between distinct perceptual experiences of the same thing, since they hold that an experience's representational content determines its phenomenal character.

just like Attenborough's. Attenborough's words fade into the background: you're aware of them in the way you'd be if you were standing next to Attenborough and hearing his narration while watching the moth. At this point, you form your beliefs about the moth on the basis of your visual experiences. If someone asks you what the moth is doing now, you don't focus on the sound of Attenborough's voice: you focus on your visual images of the moth, and answer based on those.

Yet, intuitively, you don't see the moth, or perceive that it's flitting by in the moonlight. This, despite the fact that you're having experiences just like those you'd have if you were seeing the moth, despite the fact that these experiences have the moth and its behavior among their causes, and despite the fact that the experiences represent that the moth is doing the various things it is.

Representationalists have various ways of trying to deal with this kind of case, and I don't want to suggest that they have no good answer to the Specificity Question. One possible response to the Attenborough case is that you *do* see the moth, viewing it through the medium of sound and radio waves instead of light waves. However, this is controversial, and it seems fair to say that the Specificity Question poses a non-trivial challenge to representational theories.[5]

9.2.2 Naive realism

Now let's consider a different answer to the question, "What is it, to see a red tomato?" Namely: to see a red tomato is to have a visual experience partly constituted by features of the tomato (its redness, its bulginess, etc.: whatever features of the tomato are visible to you). When you see a red tomato, the redness of your experience *is* the redness of the tomato.

This is the answer we get from naive realists about perception.[6]

Naive realism gives a clear answer to the Specificity Question. When Attenborough sees the moth, he has a visual experience with a pale silvery quality. His experience's having this quality is one and the same state of affairs as the moth's having it: the moth's instantiation of pale silveriness is one and the same instantiation of pale silveriness as the experience's instantiation of it. But even though the visual experience I have while

[5] For specificity-related criticisms of representationalism, see Searle (1983: 45–6), Robinson (1994: 165–7), and Martin (2002: 387–92).

[6] Recent defenses of naive realism include Campbell (2002), Martin (2004), Fish (2009), Brewer (2011), and Langsam (2017).

listening to Attenborough's commentary is phenomenally indistinguishable from Attenborough's experience, my experience's pale silveriness is a different instance of pale silveriness from the moth's: the quality may or may not be the same, but if the same, there are two instantiations of it: the moth's, and my experience's.[7]

Naive realism has a hard time accounting for non-veridical perceptual experiences. If I hallucinate a red tomato, the redness of my experience can't be the redness of a tomato, since there is no tomato. Most naive realists respond to this problem with a disjunctive account, according to which the redness of a veridical experience of a red tomato is a different property from the redness of a non-veridical experience of a red tomato (even though the properties are introspectively indistinguishable). Whether this is defensible is a topic of ongoing debate.[8]

Naive realism also has no easy or obvious answer to the Diversity Question. Moths, whose eyes are sensitive to ultraviolet light, see things by having visual experiences with different chromatic qualities from the experiences we humans have when seeing the same things. When Attenborough sees the moonlit moth, it's by having a visual experience with a pale silvery quality. When another moth sees the moonlit moth, it's by having a visual experience with an electric blue quality (or so let's suppose: the exact details aren't important). Both Attenborough's experience and the observing moth's are veridical. So, according to naive realism, the color of Attenborough's experience = the color of the observed moth = the color of the observing moth's experience. Since the color of Attenborough's experience ≠ the color of the observing moth's experience, this is a problem for naive realism.[9]

9.3 The phenomenalist theory

In one sentence, the phenomenalist theory of perception is that to perceive a physical thing is to have an experience that partly realizes that thing; i.e., an

[7] Naive realists often express this by saying that accurate perceptions put us into direct contact with the features of the world they give us access to, unlike non-perceptual knowledge states, which establish only an indirect connection between us and the world.

[8] For disjunctivism, see Hinton (1967), Snowdon (1980–81), Martin (2004), Fish (2009), and Langsam (2017), and for a critical assessment of disjunctivism, Byrne and Logue (2008).

[9] Some naive realists respond to this problem by saying that phenomenal differences between veridical experiences of the same thing are perceptual relations to different physical features of the beings whose experiences they are: see, e.g., Logue (2012: 222–6). Proponents of this view must take care not to identify the relevant physical features with neural representations of the percept; otherwise, their view becomes hard to distinguish from representationalism.

experience that's among those for which the thing is a possibility. How does this answer the Accuracy, Diversity, and Specificity questions?

Let's take them in reverse order.

The phenomenalist answer to the Specificity Question is that accurate perceptual experiences differ from other knowledge states by partly realizing the states of affairs of which they give us knowledge. When a blind person learns that there's a moth present through testimony from a sighted friend, his auditory experiences don't form part of the mothlike group of experiences for which the moth is a possibility; among other things, they don't occupy the same region of ideal spacetime as the experiences constituting the mothlike group.

What about the Attenborough case, where you're hearing a radio broadcast and having experiences indistinguishable from those you'd be having if standing next to the broadcaster? Here the phenomenalist diagnosis depends on whether we should classify your moth-experiences as veridical or non-veridical. Any reason to classify them as veridical is equally a reason to think that your experiences are co-located with Attenborough's in ideal spacetime, like the experiences of two people viewing the same thing from different distances, and therefore a reason to include your moth-experiences among those for which the moth is a possibility. If, on the other hand, we shouldn't count your experiences as veridical, it will be due to how the experiences relate to other experiences in the ideal manifold of possible experiences, in which case we should deny that your experiences are among those for which the moth is a possibility, on the grounds that they don't relate to other experiences (or possible experiences) in a phenomenally coherent way.

The phenomenalist answer to the Diversity Question is that two experiences are perceptions of the same thing just in case there's a physical thing such that both experiences are among those for which that thing is a possibility. When I perceive a moth visually, and a bat simultaneously perceives the same moth echolocationally, our experiences both partially realize the same physical thing: they're both among the experiences for which the moth is a possibility. Since experiences with very different phenomenal qualities can be among those for which a given physical thing is a possibility, the differences between bat and human experience are no impediment to our perceiving the same things.

Roughly, the phenomenalist answer to the Accuracy Question is that our accurate perceptual experiences are those for which various physical things are possibilities; our inaccurate perceptual experiences are those for which physical things are not possibilities.

I say "roughly," because a single experience can be accurate in some respects but inaccurate in others. The visual experience you have when you view the image on p. 91 is an example: it's a veridical experience of a checker-pattern image, but a non-veridical experience as of an image in which the region marked "A" is darker than the region marked "B." This is typical of the experiences we have when subject to optical illusions. When you see footprints as convexities on the sand, your experience is non-veridical in certain topographic respects, but veridical with respect to outline and surface area, as well as with respect to the topography of the surrounding sand; when you see a straw semi-submerged in a glass of water, your experience is non-veridical with respect to the straw's geometry (you see the straw as bent rather than straight), but veridical with respect to the straw's color, the shape of the glass, etc.; when you see a shadow cast on a white wall as a stain on the wall, your experience is non-veridical with respect to the wall's coloration, but veridical with respect to the wall's geometric features.

A more precise answer to the Accuracy Question is that an experience *e* is a veridical perception that p (where "p" designates some physical state of affairs) just in case *e* has a phenomenal feature ϕ such that *e*'s having ϕ is among the experiential states of affairs for which the fact that p is a possibility. Thus your experience of a straw semi-submerged in a glass of water is veridical with respect to the shape of the glass (since the experience's having the qualia in virtue of which it's an experience as of a glass with a certain shape is among the experiential states of affairs for which the glass's having the shape it does is a possibility), but non-veridical with respect to the shape of the straw (since the experience's having the qualia in virtue of which it's an experience as of a bent straw is not among the experiential states of affairs for which the straw's having the shape it does is a possibility).

In Chapter 2, I noted that phenomenalists owe us an account of the difference between veridical experience that's color-inverted with respect to ordinary human visual experience, and phenomenally indistinguishable but non-veridical color-inverted experience. Let me provide that account now.

First, let's get some cases on the table.

Case 1

You've taken a drug that makes orange things look blue to you; the drug has no other effects on your experience. While under the drug's influence, you look at a tangerine. Your experience of it is phenomenally blue.

Case 2
You're looking at a tangerine, having a phenomenally orange experience. At the same time, your color-inverted twin looks at the same tangerine, having a phenomenally blue experience.

Case 3
You're looking at a tangerine, having phenomenally orange experience, when your visual experience abruptly undergoes a color inversion due to a sudden rewiring of your visual cortex. Now your experience of the tangerine is phenomenally blue.

Case 4
Your bedroom ceiling is orange. While asleep one night, your visual cortex gets rewired so as to produce color-inverted experience. When you wake up and open your eyes, you have a phenomenally blue experience of your ceiling.

According to the phenomenalist theory I'm defending, a tangerine is a possibility for a hierarchy of thinglike groups of experiences (a phenomenal hypertext), and a veridical experience of a tangerine is one of the experiences belonging to the hypertext for which the tangerine is a possibility. Consider the level of the tangerine hypertext that consists of experiences like those that would occur in tangerine-observing beings capable of perceiving the microphysical surface features of physical objects (features that we, ordinary humans, can't perceive, or can perceive only with the help of powerful microscopes). Call such experiences "micro-experiences," and the level of the tangerine hypertext that includes micro-experiences the "micro-experiential subtext."

Since your experience has lower resolution than the experiences in the micro-experiential subtext, it belongs to the tangerine hypertext only if a change in your experience would be accompanied by a change in the experiences in the micro-experiential subtext. In Case 1, your experience is non-veridical (with respect to color) because it's not the case that a change in your experience's phenomenal color would be accompanied by a change in the micro-experiential subtext: if the drug wears off and your phenomenally blue experiences of the tangerine stream into phenomenally orange experiences, there will not be (or at any rate need not be) a corresponding change in the micro-experiential subtext. If a being capable of perceiving the tangerine's surface microstructure is watching the tangerine at the same time you are, that being's experiences can stream on unchanged while yours shift from blue to orange (as the drug wears off).[10]

Earlier, I argued that in scenarios like Case 2, both experiences are veridical: the difference between your visual experience and your color-inverted twin's is like the difference between equally accurate maps with inverted color schemes. The phenomenalist account of perception is consistent with this verdict. In Case 2, unlike Case 1, any change in your experience or your twin's would be accompanied by a corresponding change at the micro-experiential level (if beings with micro-experiences were looking at the tangerine at the same time). For example, if the tangerine gets painted blue while you're watching it, your experiences will go from phenomenally orange to phenomenally blue, while your twin's will go from phenomenally blue to phenomenally orange, and the experiences of micro-observers will go from experiences as of surface features that reflect light with a wavelength of around 600 nanometers (which humans normally perceive as orange) to experiences as of surface features that reflect light with a wavelength of around 480 nanometers (which humans normally perceive as blue).

In my view, your phenomenally blue experience in Case 3 is also veridical. This will strike many readers as the wrong verdict, so let me explain.

To begin with, notice that after the rewiring, you are, in effect, a color-inverted version of your previous self. Given that your color-inverted twin's experiences of the tangerine are veridical, it stands to reason that your post-rewiring experiences are veridical too.

It's true that the sudden change in your experience might make you believe, falsely, that the tangerine is now blue. If so, your phenomenally blue experiences are deceptive: deceptive, since they deceive. But veridical experiences can deceive, as when you meet someone you know, but mistake him for a stranger, due to his wearing a disguise. The phenomenally blue experiences you have immediately after your rewiring are like your experiences of your disguised friend.

There is a non-veridical experience in Case 2, but it's not the phenomenally blue experience that you have after the rewiring. The non-veridical experience is your experience as of the tangerine changing from orange to blue. This experience isn't accompanied by any corresponding change

[10] For the remainder of this discussion, by "veridical" or "non-veridical" I mean veridical or non-veridical with respect to color.

at the micro-experiential level: if micro-observers are watching the tangerine when your rewiring occurs, their experiences of the tangerine will stream on unchanged while yours shift from phenomenally orange to phenomenally blue.

Your phenomenally blue experience of the ceiling in Case 4 is also veridical, in my view. The situation here is much the same as in Case 3, only without the non-veridical experience as of a change in color. In particular, as in Case 3, there is an appropriate dependency between experiences in the micro-experiential tangerine subtext and your macro-experiences.

"But when I wake up and open my eyes, don't I experience the ceiling as having a color it really does not (namely blue)?" No. Your experience gives you a false belief about the color of the ceiling, but that doesn't prevent it from being a veridical experience. Again, the rewiring turns you into a color-inverted twin of your former self, so, given that your color-inverted twin's experiences of the ceiling are veridical, so are yours. This is something that you yourself will eventually acknowledge, if the neural rewiring persists: you'll soon stop being deceived, and learn to call objects that give you phenomenally blue experiences "orange," and ones that give you phenomenally orange experiences "blue." We might put this by saying that your phenomenally blue experiences of the ceiling are *deceptive* (at least initially) but not *defective*. They're analogous to an accurate map in an unfamiliar projection that leads you to have false beliefs about the relative sizes of various land masses.

9.4 Phenomenalists, Berkeley, and the folk

I've argued that phenomenalism has good answers to the three questions we started with. In addition, the phenomenalist theory of perception has the virtue of making the connection between veridical perceptions and their objects as close as can be, without blurring the boundary between appearance and reality. Our perceptions are intimately related to the things they're perceptions of, but that's not because the things we perceive are somehow part of our perceptions of them: it's because our perceptions are among the experiences for which the perceived things are possibilities.

To conclude this chapter, let me contrast the phenomenalist account of perception with two less widely-discussed theories that we haven't considered so far: a pre-theoretical folk theory of perception, and a Berkeleyan theory.

Here's the folk picture: there are bodies, living and nonliving, some of them ours. Our bodies interact with other bodies, and some of these interactions give rise to experiences, like the experiences I have when walking down the street. Many of these experiences have features that reflect features of the bodies that cause them, like the bodies' shapes, temperatures, and colors. We perceive things by having such experiences. As for what exactly all these bodies are, the folk picture doesn't provide any guidance on this.

Here's the Berkeleyan picture: when I walk down the street, I create the things I perceive, including the street (or parts of it), the motions of my body, and the region of space in which they occur. (I get some help from God, here.) The same goes for you and all other sentient minds: each generates a swathe of physical reality by having the perceptual experiences it does. The things we perceive *are* our perceptions of them. In Berkeley's view, perceiving things is a way of bringing them into existence.[11]

The phenomenalist picture, though in some ways reminiscent of Berkeley's and the folk's, differs from both. Physical reality is a field of phenomenal probabilities. The probabilities define a modal landscape in which experience tends to occur in ways suggestive of a physical world, given that it occurs at all. As I walk down the street, my consciousness permeates this landscape, or part of it: shaped by the relevant probabilities, my experiences realize a swathe of physical reality by being among those for which various physical things are possibilities. Likewise for other sentient beings. Neither the possibilities nor the probabilities on which they supervene depend for their existence on anyone's having any experience, so we don't create things by perceiving them. Nor is perception just the world impinging on us in felt ways. Perception is our experience being what the physical world is a possibility for. It's the world realizing itself in us.

[11] Some broadly Berkeleyan accounts of perception leave God out of the picture, and downplay the creative aspect of perception: see esp. Yetter-Chappell (2017: 71). But by all idealist accounts, our veridical perceptual experiences are parts of the physical things they're perceptions of.

10
Choose Your Own Adventure

After all is said and done, phenomenalism is a simple idea: a rock is a tendency for experiences to occur as they do when people perceive a rock, and likewise for all other physical things. I hope this idea seems less strange to you now than it may have when you began this book. I also hope I've persuaded you that, strange or not, phenomenalism has important points in its favor.

Phenomenalism is good philosophy of science: it establishes a transparent connection between the world of physics and the world of sense, it follows from a simple account of how scientific language gets its meaning, and it provides an attractive alternative to its closest rival in this area, structuralism.

Phenomenalism is good philosophy of mind: it's flexible in its approach to the mystery of consciousness, and delivers an attractive theory of perception.

Above all, phenomenalism is good metaphysics, delivering the above-listed goods by taking two things notoriously resistant to reductive analysis—chance and experience—and constructing all else out of them.

To bring these advantages into sharper focus, let's compare phenomenalism to one of the most influential and well-developed global metaphysics of the past fifty years: David Lewis's Humean supervenience. In a sound-bite, Humean supervenience

> is the doctrine that all there is to the world is a vast mosaic of local matters of particular fact, just one little thing and then another.[1]

In more detail, Humean supervenience is the view that the world fundamentally consists of elementary physical entities standing in various spatiotemporal relations. Whatever is true about our world is true in virtue of this basic physical arrangement. Some facts about our world reduce to facts about

[1] Lewis (1986a: ix). Whether "Humean supervenience" is an apt label for Lewis's position is debatable, but it's the label he favored. He worked the theory out over the whole course of his career, but he summarizes the main points in Lewis (1986a).

Phenomenalism: A Metaphysics of Chance and Experience. Michael Pelczar, Oxford University Press.
 DOI: 10.1093/oso/9780192868732.003.0010

the intrinsic nature of the arrangement; others reduce to facts about how the arrangement compares to other possible arrangements.[2]

The distinctive fundamentals of Humean supervenience are elementary physical states of affairs and "naturalness," a feature (or ostensible feature) that properties and relations have in varying degrees. All other facts reduce to, or supervene on, facts about these.[3]

Humean supervenience commits its proponents to a reductive view of consciousness, and a reductive view of chance. Lewis accepts these commitments, but he also recognizes them as his chief liabilities. He famously calls chance "the Big Bad Bug" in Humean supervenience, and his concern with consciousness is evident from his early and sustained engagement with critics of his favored brand of psychophysical reductionism.[4]

In the end, Lewis thought he could overcome all major objections to his reductionism about chance and consciousness. Be that as it may, it's easy to see why chance and consciousness seem to pose a threat to Lewis's system.

The intuitions against reductionism about chance and consciousness are very clear, and very strong. There is a possible world exactly like ours, including in all physical respects, except that people's visual experiences are color-inverted relative to ours. There is a possible world exactly like ours, including in all categorical respects, except that radon has a different half-life from what it has in our world. If facts about chance and consciousness reduced to physical or categorical facts, these worlds would be impossible. So chance and consciousness don't reduce to such facts.

Or so our intuitions tell us. Maybe our intuitions are wrong: there is an extensive literature dedicated to disparaging them. I've already suggested that I think the intuitions are sound as far as chance is concerned, and elsewhere I've argued that when it comes to consciousness, they at least have enough merit to warrant skepticism about reductive theories.[5] Here I want to make a more modest, methodological point.

The modal intuitions glossed above suggest that chance and experience are fundamental features of the world, an impression reinforced by the long history of failed attempts to give widely-accepted reductive analyses

[2] In particular, Lewis holds that facts about causation reduce to facts about how the mosaic of our world resembles the mosaics that characterize other possible worlds. Lewis (1973a) spells this out in terms of a counterfactual analysis of causation.

[3] For Lewis on naturalness, see (1983). Loosely speaking, naturalness plays a role in his metaphysics analogous to the role that coherence plays in phenomenalist and idealist metaphysics.

[4] Lewis defends an identity theory of consciousness in (1966), (1980), and (1994b). He develops his views on chance in (1986c), (1986b), and (1994a).

[5] See Pelczar (2021).

of chance and experience. Well, if we can't reduce chance and experience to anything more basic, why not reduce everything else to them? Instead of devising ever more sophisticated ways of denying what are after all very simple and natural intuitions about these things, why not take the intuitions at face value and make a virtue of necessity by making chance and experience the foundations of our metaphysics?

That's what phenomenalism does, or tries to do. It's an attempt to achieve a worldview as simple and unified as Lewis's, without claiming that consciousness reduces to something physical, or chance to something categorical.

If the attempt succeeds, it's analogous to the success of Copernican heliocentrism over Ptolemaic geocentrism. The Ptolemaic system gets a lot right when it comes to the apparent motions of heavenly bodies—so much that it's tempting to try to explain away the few details it gets wrong as due to observational errors, or minor flaws in the theory that we can fix with a suitably resourceful addition of epicycles. This, of course, is exactly what the defenders of Ptolemaic astronomy did for many generations, until people realized that by putting the Sun at the center and letting the Earth rotate, they could have a much simpler theory that gets all the details right.

Like Ptolemy's astronomy, Lewis's metaphysics gets everything right except for a few stubborn details. This makes it tempting to try to explain away the theory's problems as mere cognitive illusions, or real but minor flaws that we can fix with some resourceful ramesification.

But we should consider the possibility that these ostensibly minor flaws point to a much deeper problem with Humean supervenience: that Lewisian metaphysics gets the nature of the world fundamentally wrong. Maybe if we put experience at the center and let chance be our prime mover, we can have a much simpler metaphysics that gets all the details right. That's the possibility we've explored in this book.

At the outset I acknowledged that phenomenalism is a surprising inversion of our ordinary ways of thinking. I hope the view seems more natural to you now, and worth taking seriously. Maybe phenomenalism still isn't a path you want to go down, but at least you now have a better view of where it leads.

In responding to his critics, Mill lamented that

> men will not, and mostly cannot, throw their minds into any theory with which they are not familiar; and the bearings and consequences of the Psychological theory will have to be developed and minutely expounded

innumerable times, before it will be seen as it is, and have whatever chance it deserves of being accepted as true.[6]

There's been little expounding and even less development of phenomenalism in the sixteen decades since Mill wrote those words. If my own words have encouraged you to give phenomenalism a fair chance, they've served their purpose.

[6] Mill (1865/1979: 200–201). Mill often calls phenomenalism "the Psychological theory."

APPENDIX

Defining Spacetime Relations

In physics, to describe something as occurring at a certain time and place is to assign it (or each of its parts) a coordinate in some specified coordinate system. A coordinate is a quadruple of real-valued numbers, $\langle x, y, z, t \rangle$, and a coordinate system is an assignment of coordinates to events in which every event gets assigned a coordinate.

There are infinitely many ways to assign coordinates to physical events: infinitely many possible "coordinatizations of the manifold of physical events," as they say. You can assign events coordinates at random. You can assign the same coordinate to all events. And so forth: the possibilities are endless.

Out of all these possible coordinate systems, which ones describe physical events as occurring where and when they do in fact occur? Which systems "get it right"?

A coordinate system that gets it right—a so-called *admissible coordinate system*—is one that satisfies the following condition: when you assign physical events coordinates according to the system, you describe the events as constituting a universe in which physical events are distributed through time and space according to intelligible rules, so that, for example, later states of the universe evolve in a lawlike way from earlier states. More precisely, a coordinate system is admissible just in case it does a better job than any other system of satisfying the stated condition, where "doing a better job" is a function of how much the coordinate system permits us to account for in terms of rules that are intelligible to us, and how simple the rules are that the coordinate system facilitates. (The rules are called laws of physics.)

For example, Newton hit upon the idea of assigning coordinates to physical events in such a way that, described in terms of those coordinates, inertia is homogenous and isotropic (which is to say that the same force accelerates objects with the same mass by the same amount, no matter where, when, or in which direction the force gets applied). This is what's called an inertial coordinate system.

It turns out that when you assign coordinates to events in a way that validates the principle of inertia, you also describe the planets as obeying Kepler's Laws and, more generally, all material bodies as obeying the universal law of gravitation.

It's important to understand that these dividends of adopting an inertial coordinate system aren't logically guaranteed by the system itself. There's no contradiction in the idea of a universe in which you do *not* describe the planets as obeying Kepler's Laws when you describe the events of that universe in terms of an inertial coordinate system. Of course, Newton had a hunch that using an inertial coordinate system would have the payoffs that it turned out to have, but he couldn't have deduced those payoffs from the definition of an inertial coordinate system.

Whether a coordinate system is admissible depends on how the world is. A priori, there's no guarantee that *any* system is admissible. For all we know before looking into the matter, it might be impossible to assign coordinates to physical events in such a way that the resulting coordinate system describes a world that unfolds according to simple laws that are intelligible to us. In choosing how to assign coordinates to physical events, we're guided by a desire to assign coordinates in a way that facilitates an understanding of the world in terms of simple natural laws, and in that sense, the choice of a particular coordinate system is "conventional." But it's not up to us that any such choice is possible. There's nothing conventional about the fact that there are admissible coordinate systems, and if there weren't, there's nothing we could do about it. That our world does submit to sense-making coordinatization is, in a way, the deepest scientific mystery of all.[1]

So how do physicists assign coordinates to events? We've already seen one historic example: Newton's use of assumptions about inertia to define inertial coordinate systems. In this Appendix, we take a closer look at how contemporary physicists coordinatize the manifold of physical events. My goal here is to give a thorough inventory of the basic physical resources and methods that scientists use to describe physical things as standing in spatiotemporal relations, to make it completely clear that the phenomenal resources and methods we used to construct ideal spacetime (in Chapter 4) really are suitable for that purpose.

The search for an admissible coordinatization of physical events starts with guesswork and assumptions. Maybe you notice that certain motions or processes seem to have a periodic character, repeating themselves at regular intervals. For example, you might notice that a pendulum seems to take the same amount of time to complete each of its successive swings, and that the time always seems the same regardless of where the pendulum is located or what it's made of, provided that the length of the string is the same. Or you might notice that the Sun seems to take the same amount of time to reach its zenith from one day to the next, or that a small quartz tuning-fork seems to vibrate with a constant frequency when an electrical current passes through it, and that the frequency seems to be the same for all quartz tuning-forks of the same size and shape. Whether these seemingly periodic motions and changes really are periodic—whether they really do occur at regular intervals—is an open question at this stage of investigation. At this stage, we're just following our hunches and gut feelings about what might be a useful way to assign temporal coordinates—the t part of $\langle x, y, z, t\rangle$—to physical events.

You also notice, maybe, that some types of object seem to move, or some types of effects seem to propagate, at the same speed and without changing direction, regardless of where or when the motion or effect occurs. For example, you might notice that camel caravans travelling between Syene and Alexandria seem to advance the same distance from one sunset to the next, and conjecture that the caravans always travel at the same speed and in straight lines, regardless of where in the desert they happen to be, or in what direction they're moving. Or you might notice that a ball rolling on a smooth level surface seems to maintain the same speed and direction for most of its roll, regardless of where and in what direction the roll takes place, and that although the ball does eventually slow and stop, the time it takes it to do so is proportional to the extent to which you isolate the ball from interfering effects like air

[1] Einstein says that the everlasting mystery of the world is its comprehensibility ("Das ewig Unbegreifliche an der Welt ist ihre Begreiflichkeit": Einstein 1936: 315)

resistance and surface friction; this might lead you to speculate that objects that are entirely free from external influence move, if at all, in straight lines at constant speeds. Or you might notice that it always seems to take the same amount of time for a pulse of light bouncing between two mirrors to complete its round-trip, and conjecture that light always travels in a straight line at a constant speed. Whether any of these motions really are linear and speedwise constant is an open question at this stage. At this stage, we're just following our hunches and gut feelings about what might be a useful way to assign spatial coordinates—the x, y, and z parts of $\langle x, y, z, t\rangle$—to physical events.

The next stage is to assign coordinates to physical events in a way that's consistent with one or more of these assumptions or speculations, so that we can see if the resulting coordinate system sheds any light on what's going on in the physical world. In practice, this is the hard part—so hard that it wasn't possible to do it on a large scale until Tycho Brahe accumulated decades of precise astronomical observations, and Kepler organized them in a form that Newton could use to describe events in terms of inertial coordinates.

But these are all practical challenges. The principles behind assigning events coordinates in a way that's consistent with a given set of assumptions or speculations are simple. Here I'll describe the most up-to-date way of assigning coordinates to events: the way that does the best job (as far as we know) of facilitating a description of the physical world as unfolding according to simple natural laws. To make the presentation as clear as possible, I'm going to leave out all the guesswork, hunches, and observations that led physicists to adopt this method of coordinatizing physical events. For our purposes, the important thing is to understand the method, and get a clear view of the basic resources required to apply it. We'll take it on the scientists' authority that this method best facilitates an understanding of the physical world in terms of simple laws.[2]

First, we choose some type of physical process P in which some physical feature ϕ repeats. The plan is to use processes of this type as our chronometers. We're not taking it for granted that ϕ repeats at equal time-intervals in any given instance of the P-process, or that the time-intervals between successive iterations of ϕ in one instance of the P-process are the same as successive iterations of ϕ in other instances of the P-process. We're just identifying a certain repetitive process to see what happens when we try to use it as a clock.

It turns out that there are various physical processes that we can use as chronometers, with equally good results, but among the best processes are those that take place in atomic clocks. So, for chronometers, let's choose the type of process that takes place in the cavity of an atomic clock; the repeated feature whose successive instances serve as the "ticks" of the chronometer is the emission of an electronic impulse from the cavity.

Second, we choose some type of physical motion M that we're going to use, in combination with our chosen clocks, as a telemetric device. Again it turns out that there are various types of motion that we could use for this purpose with equally good

[2] The following discussion is adapted from Geroch (1978: 70–92).

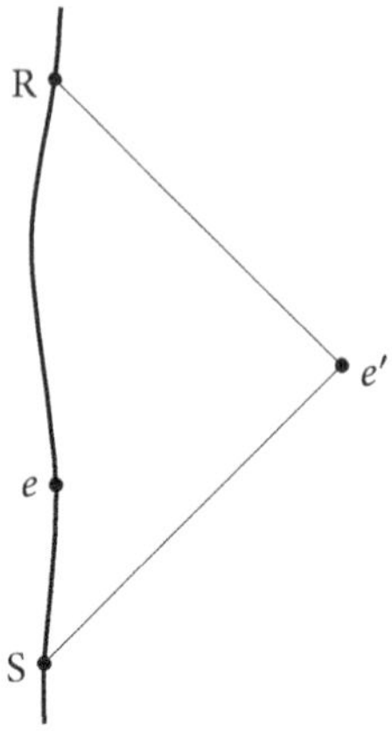

Figure A1 Assignment of spatial coordinates. S = device emits light, e' = light bounces off something, R = light strikes device, e = an event that occurs in the device

results, but among the best is the motion of light. So we'll use a combination of clocks and light signals to construct our system of coordinates.

We start by assuming that light always moves at the same speed, which we conventionally denote as c. (Whether we ultimately accept this assumption as true depends on the results we get from constructing a coordinate system based on it; for now, we're just making the assumption to see where it leads, like Newton did with his assumptions about inertial motion.)

Now, with the help of Fig. A1, we describe a method for specifying a spatial distance between any two physical events, e and e'.

The figure is a spacetime diagram. In the diagram, labeled dots represent the individual physical events we want to talk about. Lines, curved or straight, represent the worldlines of persistent physical things, comprising the events that occur in the things over the course of their existence. Dots that occur lower in the diagram represent events that occur earlier than the events represented by dots occurring higher up in the diagram. Dots occurring at a horizontal distance from one another represent events that occur at some spatial distance from one another. So, if we use a vertical line to represent an object's worldline, we represent the object as stationary; if we use a straight diagonal to represent an object's worldline, we represent the object as moving with a constant velocity; and if we use a curve to represent an object's worldline, we represent the object as accelerating. We follow the convention that a 45° line, like the one extending from S to e', and the one extending from e' to R, represents something traveling at the speed of light.

Let the bold curve in the figure represent the worldline of a device consisting of a clock attached to a light bulb. S is an event consisting of the bulb flashing (emitting a photon or light wave: you can think of "S" as standing for "send," as in "send a light signal"); e' is an event consisting of the light emitted by the flash (event S) bouncing off some object, maybe a mirror. R is an event consisting of the bounced light striking the clock-plus-bulb device (think of "R" as standing for "receive," as in "receive a light signal"). Event *e* occurs in the device sometime between events S and R. (Here we understand "between" causally: *e* occurs between S and R in the sense that S or R, but not both, causes, or contributes to the causation of, *e*.)

We now specify the spatial distance between *e* and *e*′ as one-half of the result of multiplying *c* by the difference between t_R and t_S, where t_R is the time the clock reads when *R* occurs, and t_S the time the clock reads when *S* occurs:

$$\text{Spatial distance between } e \text{ and } e' = \frac{c(t_R - t_S)}{2}$$

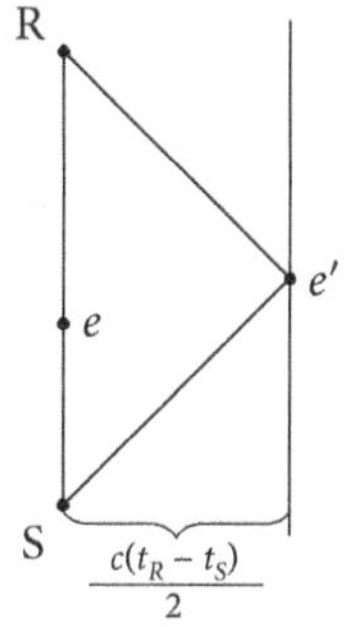

A priori, there's no guarantee that this is a good or useful way to specify the distance between the events, but it's inspired by the assumption that the speed of light is constant, and the hunch that since the light sent at S and received at R made a round trip (via the bounce at *e*′) out from and back to the device, and took t_R minus t_S seconds to make the trip, the distance between *e*′ and the device at all times between S and R is the distance that light travels in half that number of seconds. You might question this hunch: if the device moves towards or away from the mirror that bounces the light between S and R, wouldn't it be better to specify the distance between *e* and *e*′ as greater or less than the distance between S and *e*′? Surprisingly, it turns out the answer is no.[3]

Now we specify the temporal interval between events *e* and *e*′ as one half of $t_R - t_S$, minus $t_e - t_S$ (t_e is the time the clock reads when *e* occurs; see Fig. A2):

$$\text{Time elapsed between } e \text{ and } e' = \frac{t_R - t_S}{2} - (t_e - t_S)$$

Again, a priori there's no reason to think that this is a good or useful way to specify the time elapsed between the events. But it's inspired by the hunch that the light-bounce event occurred temporally halfway between events S and R, since light (we're assuming) always moves at the same speed, and so (we conjecture) took the same amount of time to move from where S happened to where *e*′ happened as it took to move from where *e*′ happened to where R happened. By the clock we're using, the amount of time that elapsed from S to the temporal halfway point between S and R is half of $t_R - t_S$. So, the time elapsed between *e* and *e*′, we surmise, is the difference between half of $t_R - t_S$ and the time elapsed between S and *e*.

Using the methods described above—"the method," for short—we can specify spatial and temporal distances between all pairs of physical events, and assign physical events spacetime coordinates in terms of which their spatial and temporal

[3] However, as explained below, ignoring the motions of the device is consistent with getting good results only if the time measured between sending and receiving events is "small," in a sense to be explained. (Notice that we're not assuming that the device is unaccelerated between S and R; from this, you can surmise that what we're constructing here isn't an inertial coordinate system.)

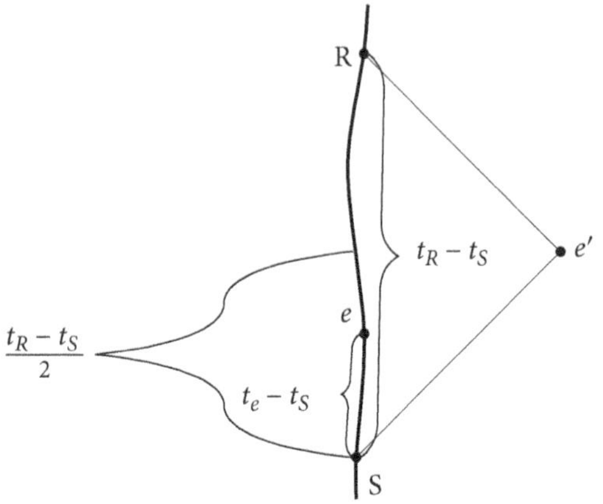

Figure A2 Assignment of temporal coordinates. $t_R - t_S$ = the difference between what the clock says when R occurs in the device and what it says when S occurs in the device. $t_e - t_S$ = the difference between what the clock says when *e* occurs in the device and what it says when S occurs in the device

distances from one another are as specified. However, it turns out that not all ways of applying the method get equally good results. You get better results by using the method only to specify spatiotemporal distances between "close" events, and then specifying distances between events that aren't close to each other by summing the distances between successive pairs of close events intermediate between the non-close ones.

We can put this more precisely. When you use the described method to specify distances between events (spatial or temporal), the chronometric worldline you use might include more or fewer sending/receiving events between a given pair of sending/receiving events. Let's say that a specification of distances that uses a chronometric worldline with more send/receive events between a given pair of send/receive events in the worldline is a "higher-resolution" specification than a specification by means of a chronometric worldline with fewer send/receive events between that pair of events (see Fig. A3).

It turns out that the higher the resolution at which you apply the method, the better the resulting coordinate system is, for the purposes of facilitating an understanding of the physical world in terms of simple laws. In the hypothetical limit, where we apply the method at infinitely high resolution, the resulting coordinate system does a perfect job of facilitating an understanding of the physical world in terms of simple laws (or at least the aspects of the physical world governed by the Einstein field equations).

Why does assigning coordinates to events this way have this useful and satisfying result? No one knows. We don't even know if there *is* a reason. For all we know, it's just a brute and inexplicable fact that this way of coordinatizing the manifold of physical events facilitates an understanding of the physical world in terms of simple

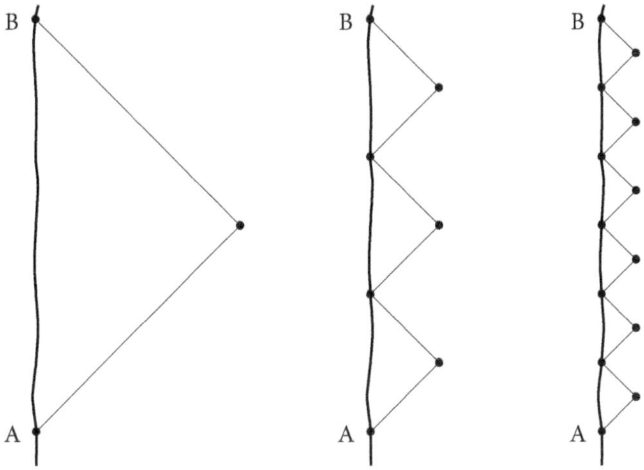

Figure A3 Coordinate system resolution. Lower resolution (left), higher resolution (middle), and still higher resolution (right). Events A and B are the same in each diagram. Curved line = worldline of a clock-and-bulb device. Dots = optical events (of light emission, reflection, or absorption)

laws. "But then how did we ever figure this out?" By a centuries-long accumulation of lucky guesses and happy intuitions borne out by laborious observation and experiment.

In reviewing how scientists define physical spacetime relations, I've skated over an issue on which scientists don't all agree, related to the distribution and causal connection of physical events in the universe. Let me briefly discuss this now.

According to our most current science, every point of spacetime contains an event, if only an event consisting of that point of spacetime having whatever properties the gravitational or electromagnetic field has at that point. Current science also tells us that every physical event belongs to worldlines—causal chains of physical events—that extend forever after that event through space, like the radii of an expanding sphere. (The causal chains might be as minimal as evolutions of the gravitational or electromagnetic field.)

If this is correct, there's a network of worldlines connecting every pair of physical events. (The connection isn't "feed forward" for every pair of events; for example, the connection between an event *y* and an event *z* might be via two worldlines that causally initiate with an event *x* and terminate with *y* and *z* respectively.) However, it's not obvious that every event belongs to a chronometric worldline. Certainly not every event occurs in a man-made clock. It may be that every physical event belongs to a chronometric worldline characterized by recurring submicroscopic events, though it's unclear that even this much is true of events occurring in the earliest epochs of the universe.[4]

[4] See Rugh and Zinkernagel (2009) and the references therein.

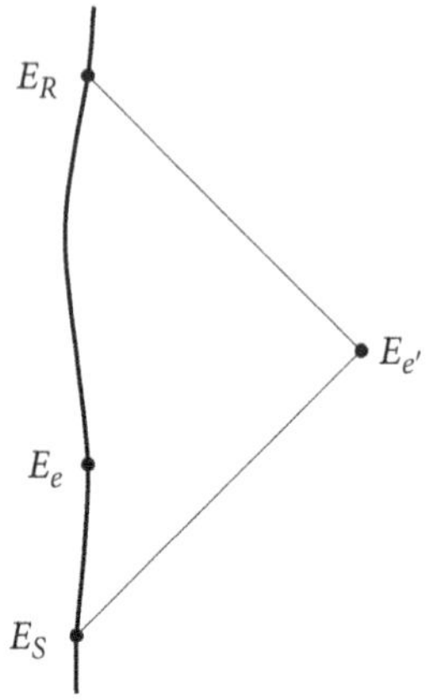

Figure A4 Assignment of coordinates in ideal spacetime. E_S, $E_{e'}$, and E_R = different experiences as of optical events, E_e = experience as of an event in a clock-and-bulb device

Even if all physical events do actually belong to chronometric worldlines intersecting with telemetric worldlines, this is at most a contingent fact about our world, and one might argue that there could be spatiotemporally related physical events in a world in which these conditions were not satisfied. There's no clear consensus among physicists whether physical events actually have to stand in physical relations like those depicted in Fig. A1 in order to stand in physical spacetime relations. According to some they do; according to others, it's enough for standing in spacetime relations if the events *would* stand in the appropriate physical relations if the world contained enough physical structure.[5]

In Chapter 4, we defined ideal spacetime relations by in effect re-interpreting Fig. A1 so that the curves and lines represent streams of consciousness, and the dots phases of those streams, as in Fig. A4. Here, the thick curved line is a stream of consciousness in which some quale recurs; this stream might consist of experiences as of a clock-and-bulb device moving through space over some period of time. E_S, E_e, and E_R are phases of this stream; E_S might be an experience as of an optical event (like a flash of light), E_e an experience as of a clock in a certain display state, and R another experience as of an optical event. Experience $E_{e'}$ might be yet another experience as of an optical event, which occurs as the last experience of a stream of consciousness that

[5] For the former view, see Eddington (1939: 70–88), Barbour (1999: 100), and Smolin (2003: 238); for the latter, see Misner et al. (1973: 814), Teller (1991), and Hoefer (1996). A third view is that physical spacetime relations are primitive and indefinable. In this view, we can't define terms like "simultaneous" or "distant from" in terms of worldlines, as above. We call on worldlines merely to fix the reference of such terms, not to give the terms their meanings, which (in this view) transcend anything we can capture with a definition in terms of actual or hypothetical causality. This view is often attributed to Newton; see also Misner (1972: 26).

begins with E_S, and the first experience of a stream that ends with E_R. The E_S-E_e-E_R stream intersects the E_S-$E_{e'}$ stream (at E_S) and the $E_{e'}$-E_R stream (at E_R).

Scientists use the method depicted in Fig. A1 to assign coordinates to physical events, and they hold that doing so allows us to locate those events in a common physical spacetime. Since the method depicted in Fig. A4 is the same method of assigning coordinates to events, only applied to experiential events instead of physical ones, we're equally entitled to hold that assigning coordinates to experiences by this method allows us to locate experiences in a common ideal spacetime. Since we can use the described method to assign coordinates to experiences occurring in different minds, the resulting spacetime is an intersubjective spacetime: a common spacetime for all experiences, not just those occurring in a single mind.

References

Adams, Robert M. 2007. "Idealism vindicated." In *Persons: Human and Divine*. Oxford: Oxford University Press, 35–54.

Albert, David Z. 2013. "Wave function realism." In *The Wave Function: Essays on the Metaphysics of Quantum Mechanics*, edited by Alyssa Ney and David Z. Albert. New York: Oxford University Press, 52–57.

Anscombe, G.E.M. 1965. "The intentionality of sensation: a grammatical feature." In *The Collected Papers of G.E.M. Anscombe*, vol. 2: *Metaphysics and the Philosophy of Mind*. Oxford: Basil Blackwell, 3–20.

Armstrong, David. 1961. *Perception and the Physical World*. London: Routledge & Kegan Paul.

Armstrong, David. 1968. *A Materialist Theory of the Mind*. London: Routledge & Kegan Paul.

Armstrong, David. 1993. "Reply to Martin." In *Ontology, Causality, and Mind: Essays in Honour of D.M. Armstrong*, edited by Keith Campbell, John Bacon, and Lloyd Reinhardt. Cambridge: Cambridge University Press, 186–94.

Armstrong, David. 1997. *A World of States of Affairs*. Cambridge: Cambridge University Press.

Armstrong, David. 2004. *Truth and Truthmakers*. Cambridge: Cambridge University Press.

Armstrong, D.M., C.B. Martin, and U.T. Place. 1996. *Dispositions: A Debate*. London: Routledge.

Ayer, A.J. 1936. *Language, Truth and Logic*. London: Victor Gollancz.

Ayer, A.J. 1946–7. "Phenomenalism." *Proceedings of the Aristotelian Society* 47: 163–96.

Ayer, A.J. 1956. *The Problem of Knowledge*. London: Macmillan.

Barbour, Julian. 1999. *The End of Time: The Next Revolution in Physics*. Oxford: Oxford University Press.

Beddor, Bob. 2020. "New work for certainty." *Philosophers' Imprint* 20(8): 1–25.

Benci, Vieri, Leon Horsten, and Sylvia Wenmackers. 2013. "Non-Archimedean probability." *Milan Journal of Mathematics* 81(1): 121–51.

Berkeley, George. 1709/1901. "An Essay Towards a New Theory of Vision." In *The Works of George Berkeley*, vol. 1, edited by A.C. Fraser. Oxford: Clarendon Press, 127–210.

Berkeley, George. 1710/1901. "A Treatise Concerning the Principles of Human Knowledge." In *The Works of George Berkeley*, vol. 1, edited by A.C. Fraser. Oxford: Clarendon Press, vol. 1, 233–347.

Berkeley, George. 1710/1982 *A Treatise Concerning the Principles of Human Knowledge*, ed. K. Winkler. Indianapolis: Hackett.

Berkeley, George. 1713/19. "Three Dialogues between Hylas and Philonous, in Opposition to Sceptics and Atheists." In *The Works of George Berkeley*, vol. 1, edited by A. C. Fraser. Oxford: Clarendon Press, 379–485.

Block, Ned. 1978. "Troubles with functionalism." *Minnesota Studies in the Philosophy of Science* 9: 261–325.

Block, Ned, and Robert Stalnaker. 1999. "Conceptual analysis, dualism, and the explanatory gap." *Philosophical Review* 108(1): 1–46.

Blume, Lawrence, Adam Brandenburger, and Eddie Dekel. 1991. "Lexicographic probabilities and choice under uncertainty." *Econometrica* 59(1): (1991) 61–79.

Bohm, David. 1952/1983. "A suggested interpretation of the quantum theory, in terms of "hidden" variables." In *Quantum Theory and Measurement*, edited by John Archibald Wheeler and Wojciech Hubert Zurek. Princeton: Princeton University Press, 369–96.

Bohr, Niels. 1928. "The quantum postulate and the recent development of atomic theory." *Nature* 121: 580–90.

Born, Max. 1854/1964. "The statistical interpretation of quantum mechanics." In *Nobel Lectures, Including Presentation Speeches and Laureates' Biographies: Physics, 1942–1962*. Amsterdam: Elsevier, 256–67.

Boscovich, Roger Joseph. 1763/1922. *A Theory of Natural Philosophy*, trans. J. M. Child. Chicago: Open Court.

Bostrom, Nick. 2003. "Are we living in a computer simulation?" *Philosophical Quarterly* 53(211): 243–55.

Boswell, James. 1799/1953. *Life of Johnson*. Oxford: Oxford University Press.

Brewer, Bill. 2011. *Perception and Its Objects*. Oxford: Oxford University Press.

Briceño, Sebastián, and Stephen Mumford. 2016. "Relations all the way down? Against ontic structural realism." In *The Metaphysics of Relations*, edited by Anna Marmodoro and David Yates. Oxford: Oxford University Press, 198–217.

Brickhill, Hazel, and Leon Horsten. 2018. "Triangulating non-Archimedean probability." *Review of Symbolic Logic* 11(3): 519–46.

Broad, C.D. 1925. *Mind and Its Place in Nature*. New York: Harcourt, Brace.

de Broglie, Louis. 1928. "La nouvelle dynamique des quanta." In *Electrons et photons: rapports et discussions du cinquième Conseil de Physique tenu à Bruxelles du 24 au 29 Octobre 1927 sous les auspices de l'Institut International de Physique Solvay*. Paris: Gauthier-Villars, 105–32.

Byrne, Alex. 2001. "Intentionalism defended." *Philosophical Review* 110(2): 199–240.

Byrne, Alex, and Heather Logue. 2008. "Either/Or." In *Disjunctivism: Perception, Action, Knowledge*, edited by Adrian Haddock and Fiona Macpherson. Oxford: Oxford University Press, 57–94.

Campbell, John. 2002. *Reference and Consciousness*. Oxford: Oxford University Press.

Campbell, Keith. 1970. *Body and Mind*. London: Macmillan.

Capaldi, Nicholas. 2004. *John Stuart Mill: A Biography*. Cambridge: Cambridge University Press.

Cargile, James. 1997. "On the burden of proof." *Philosophy* 72(279): 59–83.

Carnap, Rudolf. 1928. *Der Logische Aufbau der Welt*. Leipzig: Felix Meiner.

Chalmers, David. 1996. *The Conscious Mind: In Search of a Fundamental Theory*. Oxford: Oxford University Press.

Chalmers, David. 2010. "The Matrix as metaphysics." In *The Character of Consciousness*. Oxford: Oxford University Press, 455–94.
Chang, Hasok. 2017. "Operationalism: old lessons and new challenges." In *Reasoning in Measurement*, edited by Nicola Mößner and Alfred Nordmann. London Routledge, 25–38.
Chisholm, Roderick. 1948. "The problem of empiricism." *Journal of Philosophy* 45(19): 512–17.
Chisholm, Roderick. 1957. *Perceiving: A Philosophical Study*. Ithaca, NY: Cornell University Press.
Dainton, Barry. 2006. *Stream of Consciousness: Unity and Continuity in Conscious Experience*. New York: Routledge.
Davey, B.A., and H.A. Priestley. 2002. *Introduction to Lattices and Order*. Cambridge: Cambridge University Press.
Davidson, Donald. 2001. "A coherence theory of truth and knowledge." In *Subjective, Intersubjective, Objective*. Oxford: Clarendon Press, 137–57.
Descartes, René. 1641/1984. "Meditations on First Philosophy." In *The Philosophical Writings of Descartes*, vol. 2, edited by J. Cottingham et al. Cambridge: Cambridge University Press, 3–62.
Devitt, Michael. 1991. *Realism and Truth*. Princeton, NJ: Princeton University Press.
DeWitt, Bryce. 1970. "Quantum mechanics and reality." *Physics Today* 23(9): 30–35.
Dirac, Paul AM. 1938–9. "The relation between mathematics and physics." *Proceedings of the Royal Society of Edinburgh* 59(2): 122.
Downing, Lisa. 2001. "George Berkeley." *Stanford Encyclopedia of Philosophy, https://plato.stanford.edu/entries/berkeley/.*
Duhem, Pierre. 1914. *La Théorie physique: son objet, sa structure*. Paris: Marcel Rivière.
Eddington, A.S. 1929. *The Nature of the Physical World*. New York: Macmillan.
Eddington, A.S. 1939. *The Philosophy of Physical Science*. Cambridge: Cambridge University Press.
Einstein, Albert. 1936. "Physik und Realität." *Journal of the Franklin Institute* 221(3): 313–47.
Everett, Hugh. 1957. "'Relative state' formulation of quantum mechanics." *Reviews of Modern* 29(3): 454–62.
Feynman, Richard. 1967. *The Character of Physical Law*. Cambridge MA: MIT Press.
Fish, William. 2009. *Perception, Hallucination, and Illusion*. New York: Oxford University Press.
Foster, John. 1982. *The Case for Idealism*. London: Routledge & Kegan Paul.
Foster, John. 1991. *The Immaterial Self*. London: Routledge.
Foster, John. 1993. "The succinct case for idealism." In *Objections to Physicalism*, edited by Howard Robinson. Oxford: Clarendon Press, 293–313.
Foster, John. 2008. *A World for Us: The Case for Phenomenalistic Idealism*. Oxford: Oxford University Press.
van Fraassen, Bas. 1980. *The Scientific Image*. Oxford: Oxford University Press.
French, Steven. 2014. *The Structure of the World: Metaphysics and Representation*. Oxford: Oxford University Press.

Fumerton, Richard. 1985. *Metaphysical and Epistemological Problems of Perception*. Lincoln: University of Nebraska Press.

Gauss, Carl Friedrich. 1827/1902. *General Investigations of Curved Surfaces of 1827 and 1825*. Princeton, NJ: Princeton University Library.

Geroch, Robert. 1978. *General Relativity from A to B*. Chicago: University of Chicago Press.

Goff, Philip. 2017. *Consciousness and Fundamental Reality*. New York: Oxford University Press.

Grote, John. 1865. *Exploratorio Philosophica: Rough Notes on Modern Intellectual Science*. Cambridge: Deighton, Bell.

Hájek, Alan. 2014a. *Most Counterfactuals are False*. https://philarchive.org/archive/HJEMCAv1.

Hájek, Alan. 2014b. "Probabilities of counterfactuals and counterfactual probabilities." *Journal of Applied Logic* 12(3): 235–51.

Halpern, Joseph Y. 2010. "Lexicographic probability, conditional probability, and nonstandard probability." *Games and Economic Behavior* 68(1): 155–79.

Hamilton, Andy. 1998. "Mill, phenomenalism, and the self." In *The Cambridge Companion to Mill*, edited by John Skorupski. Cambridge: Cambridge University Press, 139–75.

Hammond, Peter J. 1994. "Elementary non-Archimedean representations of probability for decision theory and games." In *Patrick Suppes: Scientific Philosopher*, edited by Paul Humphreys. Dordrecht: Kluwer Academic, 25–59.

Harman, Gilbert. 1990. "The intrinsic quality of experience." *Philosophical Perspectives* 4: 31–52.

Hartshorne, Charles. 1946. "Leibniz's greatest discovery." *Journal of the History of Ideas* 7(4): 411–21.

Heisenberg, Werner. 1927/1983. "The physical content of quantum kinematics and mechanics." In *Quantum Theory and Measurement*, edited by John Archibald Wheeler and Wojciech Hubert Zurek. Princeton, NJ: Princeton University Press, 62–84.

Hinton, J.M. 1967. "Visual experiences." *Mind* 76(302): 217–27.

Hobbes, Thomas. 1655/1839. *Concerning Body*. London: John Bohn.

Hoefer, Carl. 1996. "The metaphysics of space-time substantivalism." *Journal of Philosophy* 93(1): 5–27.

Huemer, Michael. 2015. "Serious theories and skeptical theories: why you are probably not a brain in a vat." *Philosophical Studies* 173(4): 1031–52.

Hume, David. 1739/2007. *A Treatise of Human Nature*. Oxford: Clarendon Press.

Jackson, Frank. 1977. *Perception: A Representative Theory*. Cambridge: Cambridge University Press.

Jackson, Frank. 1982. "Epiphenomenal qualia." *Philosophical Quarterly* 32(127): 23–36.

Jevons, W. Stanley. 1875. *Money and the Mechanism of Exchange*. New York: D. Appleton.

Jolley, Nicholas. 2005. *Leibniz*. London: Routledge.

Kant, Immanuel. 1781/1998. *Critique of Pure Reason*. Cambridge: Cambridge University Press.

Kant, Immanuel. 1783/1997. *Prolegomena to Any Future Metaphysics That Will Be Able to Come Forward as Science*. Cambridge: Cambridge University Press.

Kirk, Robert. 1974. "Sentience and behavior." *Mind* 83(329): 43–60.

Kiyotaki, Nobuhiro, and Randall Wright. 1991. "A contribution to the pure theory of money." *Journal of Economic Theory* 53(2): 215–35.

Kriegel, Uriah. 2009. *Subjective Consciousness: A Self-Representational Theory*. Oxford: Oxford University Press.

Kripke, Saul. 1980. *Naming and Necessity*, revised edn. Oxford: Basil Blackwell.

Ladyman, James, and Don Ross. 2007. "Ontic structural realism and the philosophy of physics." In *Every Thing Must Go: Metaphysics Naturalized*, edited by James Ladyman et al. New York: Oxford University Press, 130–89.

Ladyman, James, Don Ross, David Spurret, and John Collier, eds. 2007. *Every Thing Must Go: Metaphysics Naturalized*. Oxford: Oxford University Press.

Langsam, Harold. 2017. "The intuitive case for naive realism." *Philosophical Explorations* 20(1): 106–22.

Langsam, Harold. 2020. "Why intentionalism cannot explain phenomenal character." *Erkenntnis* 85(2): 375–89.

Langton, Rae. 1998. *Kantian Humility: Our Ignorance of Things in Themselves*. Oxford: Clarendon Press.

Leibniz, G.W. 1704/2013. "Leibniz to de Volder, 30 June 1704." In *The Leibniz–De Volder Correspondence*, edited by Paul Lodge. New Haven, CT: Yale University Press, 301–10.

Leibniz, G.W. 1705/2013. "Leibniz to de Volder, January 1707?" In *The Leibniz–De Volder Correspondence*, edited by Paul Lodge. New Haven, CT: Yale University Press, 319–24.

Leibniz, G.W. 1712/1989. "Notes for Leibniz to Des Bosses, 5 February 1712." In *Philosophical Essays*, edited by Roger Ariew and Daniel Garber. Indianapolis: Hackett, 199–200.

Leibniz, G.W. 1712/2007a. "Leibniz to Des Bosses, 15 February 1712." In *The Leibniz–Des Bosses Correspondence*, edited by Brandon C. Look and Donald Rutherford. New Haven CT: Yale University Press, 221–35.

Leibniz, G.W. 1712/2007b. "Leibniz to Des Bosses, 16 June 1712." In *The Leibniz–Des Bosses Correspondence*, edited by Brandon C. Look and Donald Rutherford. New Haven, CT: Yale University Press, 254–7.

Leibniz, G.W. 1712/2007c. "Leibniz to Des Bosses, 19 August 1712." In *The Leibniz–Des Bosses Correspondence*, edited by Brandon C. Look and Donald Rutherford. New Haven, CT: Yale University Press, 346–53.

Leibniz, G.W. 1712/2007d. "Leibniz to Des Bosses, 24 January 1713." In *The Leibniz–Des Bosses Correspondence*, edited by Brandon C. Look and Donald Rutherford. New Haven, CT: Yale University Press, 294–313.

Leibniz, G.W. 1712/2007e. "Leibniz to Des Bosses, 26 May 1712." In *The Leibniz–Des Bosses Correspondence*, edited by Brandon C. Look and Donald Rutherford, New Haven, CT: Yale University Press, 240–45.

Leibniz, G.W. 1714. "Principles of Nature and Grace, Based on Reason." In *Philosophical Essays*, edited by Roger Ariew and Daniel Garber. Indianapolis: Hackett, 206–13.

Leibniz, G.W. 1714/1989. "Monadology." In *Philosophical Essays*, edited by Roger Ariew and Daniel Garber. Indianapolis: Hackett 213–25.

Leibniz, G.W. 1715/1989. "Conversation of Philarète and Ariste." In *Philosophical Essays*, edited by Roger Ariew and Daniel Garber. Indianapoli: Hackett 257–68.

Leibniz, G.W. 1998. *G.W. Leibniz: Philosophical Texts*. Oxford: Oxford University Press.

Leibniz, G.W. 2007. *The Leibniz-Des Bosses Correspondence* edited by Brandon C. Look and Donald Rutherford. New Haven, CT: Yale University Press.

Leitgeb, Hannes. 2012a. "A probabilistic semantics for counterfactuals: part A." *Review of Symbolic Logic* 5(1): 26–84.

Leitgeb, Hannes. 2012b. "A probabilistic semantics for counterfactuals: part B." *Review of Symbolic Logic* 5(2): 85–121.

Lewis, Clarence Irving. 1946. *An Analysis of Knowledge and Valuation*. La Salle, IL: Open Court.

Lewis, David. 1966. "An argument for the identity theory." *Journal of Philosophy* 63(1): 17–25.

Lewis, David. 1972. "Psychophysical and theoretical identifications." *Australasian Journal of Philosophy* 50(3): 249–58.

Lewis, David. 1973a. "Causation." *Journal of Philosophy* 70(17): 556–67.

Lewis, David. 1973b. *Counterfactuals*. Malden, MA: Basil Blackwell.

Lewis, David. 1980. "Mad pain and Martian pain." In *Readings in Philosophy of Psychology*, vol. 1, edited by Ned Block. Cambridge, MA: Harvard University Press, 216–22.

Lewis, David. 1981. "A subjectivist's guide to objective chance." In *IFS: Conditionals, Belief, Decision, Chance, and Time*, edited by William L. Harper, Robert Stalnaker, and Glenn Pearce. Dordrecht: D. Reidel, 267–97.

Lewis, David. 1983. "New work for a theory of universals." *Australasian Journal of Philosophy* 61(4): 343–77.

Lewis, David. 1986. "Introduction." In *Philosophical Papers*, vol. 2. Oxford: Oxford University Press, ix–xvii.

Lewis, David. 1986b. "Probabilities of conditionals and conditional probabilities." In *Philosophical Papers*, vol. 2. Oxford: Oxford University Press, 133–56.

Lewis, David. 1986c. "A subjectivist's guide to objective chance." In *Philosophical Papers*, vol. 2. Oxford: Oxford University Press, 83–132.

Lewis, David. 1994a. "Humean supervenience debugged." *Mind* 103(412): 473–90.

Lewis, David. 1994b. "Reduction of mind." In *A Companion to Philosophy of Mind*, edited by Samuel Guttenplan. Oxford: Blackwell, 412–31.

Loar, Brian. 2003. "Transparent experience and the availability of qualia." In *Consciousness: New Philosophical Perspectives*, edited by Quentin Smith and Aleksandar Jokic Oxford: Oxford University Press, 77–96.

Locke, John. 1694/1979. *An Essay Concerning Human Understanding*. Oxford: Clarendon Press.

Logue, Heather. 2012. "Why naive realism?" *Proceedings of the Aristotelian Society* 112: 211–37.

Mabbott, J.D. 1931. "The place of God in Berkeley's philosophy." *Journal of Philosophical Studies* 6(21): 18–29.

Mackie, J.L. 1976. *Problems from Locke*. Oxford: Clarendon Press.

Macleod, Christopher. 2020 "John Stuart Mill." *Stanford Encyclopedia of Philosophy*, https://plato.stanford.edu/entries/mill/, accessed September 28, 2020.

Martin, M.G.F. 2002. "The transparency of experience." *Mind & Language* 17(4): 376–425.

Martin, M.G.F. 2004. "The limits of self-awareness." *Philosophical Studies* 120(1/3): 37–89.

Maxwell, Nicholas. 1966. "Physics and common sense." *British Journal for the Philosophy of Science* 16(64): 295–311.

McDermott, Michael. 1999. "Counterfactuals and access points." *Mind* 108(430): 291–334.

McGee, Vann. 1994. "Learning the impossible." In *Probability and Conditionals: Belief Revision and Rational Decision*, edited by Ellery Eells and Brian Skyrms. Cambridge: Cambridge University Press, 179–199.

McKitrick, Jennifer. 2003. "The bare metaphysical possibility of bare dispositions." *Philosophy and Phenomenological Research* 66(2): 349–69.

Merricks, Trenton. 2001. *Objects and Persons*. Oxford: Oxford University Press.

Mill, John Stuart. 1843/1974. *A System of Logic Ratiocinative and Inductive: Being a Connected View of the Principles of Evidence and the Methods of Scientific Investigation*. Toronto: University of Toronto Press.

Mill, John Stuart. 1865/1972. "Letter to Herbert Spencer, 12 August 1865." In *The Later Letters of John Stuart Mill: 1949–1873*. Toronto University of Toronto Press, 1089–91.

Mill, John Stuart. 1865/1979. *An Examination of Sir William Hamilton's Philosophy, and of the Principal Philosophical Questions Discussed in his Writings*. Toronto: University of Toronto Press.

Mill, John Stuart. 1871/1978. "Berkeley's life and writings." In *The Collected Works of John Stuart Mill*, vol. 11: *Essays on Philosophy and the Classics*, edited by John M. Robson. Toronto: University of Toronto Press, 451–71.

Millar, Boyd. 2014. "The phenomenological directness of perceptual experience." *Philosophical Studies* 170(2): 235–53.

Misner, Charles W. 1972. "Some topics for philosophical inquiry concerning the theories of mathematical geometrodynamics and of physical geometrodynamics." *PSA: Proceedings of the Biennial Meeting of the Philosophy of Science Association*, 7–29.

Misner, Charles W., Kip S. Thorne, and John Archibald Wheeler. 1973. *Gravitation*. San Francisco, CA: W.H. Freeman.

Monton, Bradley. 2013. "Against 3N-dimensional space." In *The Wave Function: Essays on the Metaphysics of Quantum Mechanics*, edited by Alyssa Ney and David Z. Albert. New York: Oxford University Press, 154–67.

Moore, G.E. 1939. "Proof of an external world." *Proceedings of the British Academy* 25: 273–300.

Mumford, Stephen. 2006. "The ungrounded argument." *Synthese* 149(3): 471–89.

Nagel, Thomas. 1974. "What is it like to be a bat?" *Philosophical Review* 83(4): 435–50.

Newman, M.H.A. 1928. "Mr. Russell's 'Causal Theory of Perception'." *Mind* 37(146): 137–48.

Nisbett, Richard, and Lee Ross. 1980. *Human Inference: Strategies and Shortcomings of Social Judgment*. Englewood Cliffs, NJ: Prentice-Hall.

Parfit, Derek. 1984. *Reasons and Persons*. Oxford: Clarendon Press.

Pautz, Adam. 2010. "Why explain visual experience in terms of content?" In *Perceiving the World*, edited by Bence Nanay. Oxford: Oxford University Press, 254–309.

Pelczar, Michael. 2010. "Must an appearance of succession involve a succession of appearances?" *Philosophy and Phenomenological Research* 81(1): 49–63.

Pelczar, Michael. 2015. *Sensorama: A Phenomenalist Analysis of Spacetime and Its Contents*. Oxford: Oxford University Press.

Pelczar, Michael. 2019. "Defending phenomenalism." *Philosophical Quarterly* 69(276): 574–97.

Pelczar, Michael. 2021. "Modal arguments against materialism." *Noûs* 55(2): 426–44.

Pitcher, George. 1971. *A Theory of Perception*. Princeton, NJ: Princeton University Press.

Place, U.T. 1956. "Is consciousness a brain process?" *British Journal of Psychology* 47: 44–50.

Poincaré, Henri. 1898. "La mesure du temps." *Revue de métaphysique et de morale* 6: 1–13.

Popper, Karl. 1959/1992. *The Logic of Scientific Discovery*. London: Routledge.

Popper, Karl. 1983. *Realism and the Aim of Science*. London: Routledge.

Price, H.H. 1932. *Perception*. London: Methuen.

Price, H.H. 1940. *Hume's Theory of the External World*. Oxford: Clarendon Press.

Putnam, Hilary. 1982. "Brains in a vat." In *Reason, Truth, and History*. Cambridge: Cambridge University Press, 1–21.

Quine, Willard Van Orman. 1948/1961. "On what there is." In *From a Logical Point of View*. Cambridge, MA: Harvard University Press, 1–19.

Quine, Willard Van Orman. 1981. "Things and their place in theories." In *Theories and Things*. Cambridge, MA: Harvard University Press, 1–23.

Robinson, Howard. 1982. *Matter and Sense: A Critique of Contemporary Materialism*. Cambridge: Cambridge University Press.

Robinson, Howard. 1994. *Perception*. London: Routledge.

Robinson, Howard. Forthcoming. *Perception and Idealism*. Oxford: Oxford University Press.

Roeper, Peter, and Hugues Leblanc. 1999. *Probability Theory and Probability Logic*. Toronto: Toronto University Press.

Rugh, Svend E., and Henrik Zinkernagel. 2009. "On the physical basis of cosmic time." *Studies in History and Philosophy of Science Part B: Studies in History and Philosophy of Modern Physics* 40(1): 1–19.

Russell, Bertrand. 1912. *The Problems of Philosophy*. London: Williams & Norgate.

Russell, Bertrand. 1918. "The philosophy of logical atomism." *The Monist* 28: 495–527.

Russell, Bertrand. 1926. "The world of physics and the world of sense." In *Our Knowledge of the External World as a Field for Scientific Method in Philosophy*, 2nd edn. London: George Allen & Unwin, 106–34.

Russell, Bertrand. 1927. *The Analysis of Matter*. London: Kegan Paul, Trench, Trubner.

Saunders, Simon. 2003. "Structural realism, again." *Synthese* 136(1): 127–33.
Scarre, Geoffrey. 1980. *Logic and Reality in the Philosophy of John Stuart Mill.* Dordrecht: Kluwer Academic.
Schoeman, Ferdinand. 1987. "Statistical vs. direct evidence." *Noûs* 21(2): 179–98.
Schrödinger, Erwin. 1935/1983. "The present situation in quantum mechanics." In *Quantum Theory and Measurement*, edited by John Archibald Wheeler and Wojciech Hubert Zurek. Princeton, NJ: Princeton University Press, 152–67.
Searle, John. 1983. *Intentionality: An Essay in the Philosophy of Mind.* Cambridge: Cambridge University Press.
Sellars, Wilfred. 1962. "Philosophy and the scientific image of man." In *Frontiers of Science and Philosophy*. Pittsburgh, PA: University of Pittsburgh Press, 37–78.
Shoemaker, Sydney. 1994. "Phenomenal character." *Noûs* 28(1): 21–38.
Sider, Theodore. 2011 *Writing the Book of the World*. Oxford: Clarendon Press.
Skorupski, John. 1989. *John Stuart Mill.* London: Routledge.
Skorupski, John. 1994. "J.S. Mill: logic and metaphysics." In *The Nineteenth Century*, edited by C.L. Ten. London: Routledge, 98–121.
Smart, J.J.C. 1959. "Sensations and brain processes." *Philosophical Review* 68(2): 141–56.
Smart, J.J.C. 1963. *Philosophy and Scientific Realism*. London: Routledge & Kegan Paul.
Smithson, Robert. 2017. "A new epistemic argument for idealism." In *Idealism: New Essays in Metaphysics*, edited by Tyron Goldschmidt and Kenneth L. Pearce. Oxford: Oxford University Press, 17–33.
Smolin, Lee. 2003. "Time, structure and evolution in cosmology." In *Revisiting the Foundations of Relativistic Physics: Festschrift in Honor of John Stachel*, edited by Jürgen Renn, Lindy Divarci, and Petra Schröter. Dordrecht: Kluwer Academic, 221–74.
Snowdon, Paul. 1980–81. "Vision and causation." *Proceedings of the Aristotelian Society* 81: 175–92.
Spinoza, Baruch. 1677/1985. "Ethics." In *The Collected Works of Spinoza*, vol. I, edited by Edwin Curley. Princeton, NJ: Princeton University Press, 213–382.
Stalnaker, Robert C. 1968. "A theory of conditionals." In *Studies in Logical Theory*, edited by Nicholas Rescher. Oxford: Basil Blackwell, 98–112.
Stalnaker, Robert C. 1970. "Probability and conditionals." *Philosophy of Science* 37(1): 64–80.
Stoljar, Daniel, and Michael Smith. 1998. "Global response-dependence and noumenal realism." *The Monist* 81(1): 85–111.
Strawson, Galen. 2006. "Realistic monism: why physicalism entails panpsychism." In *Consciousness and Its Place in Nature: Does Physicalism Entail Panpsychism?*, edited by Anthony Freeman. Exeter: Imprint Academic, 3–29.
Strawson, Galen. 2008. *Real Materialism: and Other Essays*. Oxford: Clarendon Press.
Tegmark, Max. 2014. *Our Mathematical Universe: My Quest for the Ultimate Nature of Reality*. New York: Alfred A. Knopf.
Teller, Paul. 1991. "Substance, relations, and arguments about the nature of space-time." *Philosophical Review* 100(3): 363–97.

Thomasson, Amie L. 2007. *Ordinary Objects*. Oxford: Oxford University Press.

Tye, Michael. 1995. *Ten Problems of Consciousness*. Cambridge, MA: MIT Press.

Vogel, Jonathan. 1990. "Cartesian skepticism and inference to the best explanation." *Journal of Philosophy* 87(11): 658–66.

Vogel, Jonathan. 2005. "The refutation of skepticism." In *Contemporary Debates in Epistemology*, edited by Matthias Steup and Ernest Sosa. Malden, MA: Blackwell, 72–84.

Warnock, G.J. 1969. *Berkeley*. Harmondsworth: Penguin (Peregrine Books).

Winkler, Kenneth P. 1989. *Berkeley: An Interpretation*. Oxford: Clarendon Press.

Yetter-Chappell, Helen. 2017. "Idealism without God." In *Idealism: New Essays in Metaphysics*, edited by Tyron Goldschmidt and Kenneth L. Pearce. Oxford: Oxford University Press, 66–81.

Index of Names

Subject Index